Oil: Prices and Capital

Oil: Prices and Capital

Haim Ben-Shahar
Tel-Aviv University

Lexington Books
D.C. Heath and Company
Lexington, Massachusetts
Toronto

Library of Congress Cataloging in Publication Data

Ben-Shahar, Haim.
   Oil: prices and capital.

   Bibliography: p.
   1. Petroleum products—Prices. 2. Price policy.
3. Petroleum industry and trade—Near East—Finance.
I. Title.
HD9560.4.B43   338.2'3   75-5239
ISBN 0-669-99804-4

Published simultaneously in Canada.

Printed in the United States of America.

International Standard Book Number: 0-669-99804-4.

Library of Congress Catalog Card Number: 75-5239.

Contents

viii

List of Figures

List of Tables

Acknowledgments

The research for this book started in 1974 while I was on sabbatical leave at Columbia University, Graduate School of Business Administration. The book is partly based on an earlier study made at the Hudson Institute in summer, 1974. The work was completed at the Foerder Institute of Economic Research at Tel-Aviv University. I am deeply indebted to these institutions, for their financial help and services without which this book could not have been written.

The work was greatly stimulated by Herman Kahn, Robert Schatz, and their staff at the Hudson Institute, to whom I owe special thanks.

My devoted research assistants at Columbia University and at Tel-Aviv University should be credited for more than I can actually express. In particular I would like to express my thanks to Charles Boynton, Jean-Pierre Jordan, Bharat Shrestha, Vance McCracken, and to Yehezkel Besalel, Avi Harel, Mark Leidig, and Dan Levin.

I would like to extend my thanks to Frank Werner, Shaul Beit-Aa'ron Jacob Shaya, and Michael Komei who developed the computer programs of the models and thereby provided invaluable services. I am indebted to Carol Gillis, Stella Fedida, Rosette Kryger, and Helen Miller who typed the several drafts, and to Lew Golan who edited the final draft. I wish to express a particular indebtedness to Uzi Arad who helped me in all stages of the study by his sound advice and general services. Last, but not least, Professors Eytan Berglas and Robert Hawkins read an earlier draft and provided invaluable comments.

It is impossible to mention all my colleagues who have contributed directly or indirectly to the study and to whom I owe great thanks. Only for the errors do I take full responsibility.

Introduction

The 1973 energy crisis caught the world unprepared — both practically and psychologically; thus its effect was extremely powerful, and the embargo created near panic. Political and economic thought was also unprepared. Political concepts were still in the 1960s, namely, that the market was dominated by strong oil companies and that the governments of the oil producing countries were no stronger (and usually weaker) than the oil companies. Although economic research on specific and technical aspects of oil economy was befitting such an important branch of world economics, theoretical economic conceptualization lagged behind. This was indicated by a relative lack of interest in the economics of exhaustible assets and its application to energy.

The outbreak of the energy crisis spurred the application of economic thought to the oil industry. Research on the theory of exhaustible assets became intense. Attempts to apply price theory under uncertainty to the energy industry have been made. But most of the enormous flow of publications during the first two years of the crisis was commentative rather than analytical.

The decision to write this book followed the preliminary findings of a draft monograph prepared by the author for the Hudson Institute during the early months of the crisis. That study showed that theory and analytical techniques could clarify the economic aspects of the energy problems, thereby shedding some light on related political factors as well. Therefore, this book is based on analytical methods that are well known in economics but not widely applied to the oil industry.

While the approach is analytic, it is oriented toward application. Moreover, the subject matter is presented in popular fashion; no familiarity with relevant economic theory is required, so it can be read easily by noneconomists.

The book deals with three areas: (1) energy prices and price policy; (2) economic growth and foreign capital accumulation of the main oil-producing countries; and (3) investment policy for foreign capital.

Part 1 deals with energy prices and price policy. Chapter 1 gives a background description of the oil industry; Chapter 2 presents the methodology. Chapters 3 and 4 describe the world supply and demand for energy and oil. In Chapter 5 these data are incorporated into a dynamic programming model designed to optimize price policy. The following two chapters apply this model to the oil market. Chapter 6 analyzes the price policy of the Organization of Petroleum Exporting Countries (OPEC) as a cohesive cartel behaving as a price-leader oligopoly in order to explain past price behavior (in particular, the 1973-1974 hike) and discuss future price developments. Chapter 7 applies the same analytical tools to separate subgroups within OPEC—revealing which countries benefit more than others from the present OPEC price policy and the conflicts of interest among them. These conflicts constitute a source of political hypotheses and a tool for political planning that can be used by both consuming and producing countries.

Part 2 deals with the economic development of the main oil-producing countries within OPEC (the Middle East countries) and the relationship between their oil revenues and economic development. It presents an optimization framework for explaining the allocation of oil revenues for domestic uses (consumption and investments) and foreign investments using a simple economic growth model. Oil revenues are treated as a separate factor of the economic resources. Funds are allocated to domestic and foreign investments in a manner that maximizes their total contribution to the gross national product. Chapter 8 presents the conceptual framework for Part 2. Chapter 9 describes the model of economic growth, allocation of investments, and capital accumulation. Chapter 10 summarizes the main projections of economic development, consumption, investments, foreign trade, and accumulation of foreign capital. These results are presented for alternative assumptions regarding future oil prices and other economic parameters. They are compared with other published projections thereby revealing the implicit assumptions behind different forecasts.

Part 3 (Chapters 11 and 12) deals with investment policies for the foreign capital accumulated by the OPEC members. The discussion, again, is conceptual and analytic, applying a model for optimal allocation of capital to various available Western investment outlets. The model is based on the preferential criteria of investors and the contribution of each available investment tool to each criterion. The optimal allocation is that which maximizes the level of achievement of the investors' preferences. A specific exercise clarifies the model mechanism (Chapter 12).

I regard this book as a modest step toward better understanding of the energy crisis. It is based on economic theory and applies available analytical tools presented in an elementary fashion. This makes the book suitable for any intelligent reader interested in the subject.

Part 1

The Policy of Oil Price Determination

1

The Oil Market Structure

Historical Development

Since World War II there has been a rapid and consistent growth in the production and consumption of energy. The main forms of energy now in use are oil, coal, and gas. Total energy output in 1948 was the equivalent of 13 bil.bbl. of oil.[1] This increased to 42 bil.bbl. by 1973,[2] an average yearly increase of 5 percent. The world's per-capita consumption of energy increased from the equivalent of 6 bbl. of oil in 1950 to over 11 bbl. in 1973—a 2.8 percent annual increase.

The distribution of this consumption was not even. Energy consumption in underdeveloped countries has been insignificant. Industrially developed countries consume almost 90 percent of the world's energy output. The highest level of consumption is in the United States—the equivalent of over 60 bbl. of oil per capita per annum, almost 10 times the world-wide average.

The most significant development in the postwar energy market has been the increasing market share of oil in place of coal. In 1950 coal accounted for 62 percent of the world's energy; in 1973 it accounted for only 30 percent. Simultaneously, oil's market share doubled from 25 percent to almost 50 percent.

Table 1-1 shows the tremendous increase in oil output—from 3.6 bil.bbl. in 1950 to 20.2 bil.bbl. in 1973—an annual increase of 7.6 percent. The

Table 1-1
Energy Output by Sources, 1950-1973

	1950		1960		1970		1973	
	Bil. Bbl. Equiv.	Percent	Bil. Bbl. Equiv.	Percent	Bil. Bbl. Equiv.	Percent	Bil. Bbl. Equiv.	Percent
Oil	3.6	25.5	7.4	31.2	16.0	41.9	20.2	48.5
Coal	8.8	62.4	12.3	51.9	13.4	35.1	12.7	30.7
Gas	1.5	10.6	3.5	14.8	7.9	20.7	7.7	18.5
Other	0.2	1.5	0.5	2.1	0.9	2.3	1.0	2.3
Total	14.1	100.0	23.7	100.0	38.2	100.0	41.6	100.0

Sources: 1950: *World Energy Supply, 1956-1959*, series J, no. 4, New York, 1961, pp. 8-9; 1960: *W.E.S.,1960-1963*, series J, no. 8, 1965, pp. 10-11; 1970: *W.E.S., 1968-1971*, series J, no. 16, 1973, pp. 5-6; 1973: *B.P. Statistical Review of the World Oil Industry, 1973*, London, 1974.

3

increasing importance of natural gas can also be seen; its market share doubled from 10 percent in 1950 to approximately 20 percent in 1973. The shares of other energy sources have been relatively insignificant.

The increases in energy consumption and in the market share of oil both resulted from a long-term decline in the price of oil. In 1959 oil in Western Europe cost $3/bbl.[3] Thereafter, the price declined steadily, reaching a low of $2/bbl. in 1966. This downward trend ended in 1967, but the price remained low—around $2.30/bbl.—until 1970. In the early 1970s the price increased—reaching about $3/bbl. in October 1973. Considering world inflation, the 1973 price in real terms was only about half the 1959 price. This oil price decline in real terms encouraged an increase in energy demand. Furthermore, there was a significant decrease in the energy unit price of oil relative to coal. This trend in Great Britain is shown in Table 1-2.

The price of oil relative to coal in Great Britain declined by 46 percent from 1954 to 1965. Even after the trend in oil prices changed, this ratio was still 38 percent lower in 1973 than in 1954. This price development provides a major explanation for the significant shift in consumption from coal to oil, which was further encouraged by the greater convenience of using oil.

The Oil Market

The structure of the oil market is determined primarily by the availability and distribution of oil reserves. Total world reserves amounted to 720 bil.bbl. in 1974. Approximately 70 percent of the reserves belong to members of the Organization of Petroleum Exporting Countries (OPEC), and another 15 percent to communist countries; the non-OPEC western countries have only 10 percent. The United States has 35 bil.bbl., Western Europe 25 bil.bbl., and

Table 1-2
Prices of Oil and Coal in Great Britain, 1954-1973
(£ per Ton Oil Equivalent)

Year	Oil	Coal	Ratio of Oil/Coal
1954	8.60	5.20	1.51
1960	8.30	8.10	1.02
1965	7.00	8.70	0.80
1970	9.30	10.00	0.93
1973	12.20	13.00	0.94

Source: *United Kingdom Energy Statistics*, Dept. of Trade and Industry, London, 1973, p. 139.

Canada 10 bil.bbl. The available reserves are sufficient for only 20 years if production continues to increase at the same rate as in the past. The uneven distribution of reserves intensifies the problem.

Historically the oil market has been controlled by a relatively small number of big oil companies. These companies possessed a significant degree of monopolistic power, since they held a large share of the world market and cooperated in their international activities. With the postwar increase in oil consumption, the companies shifted the center of their operations from their home countries to areas where new oil reserves were being discovered—especially the Middle East. Consequently, the Middle East's share in world oil output increased from 15 percent in 1950 to over 40 percent in 1973.

In 1960 OPEC was established; it includes Saudi Arabia, Kuwait, United Arab Emirates, Iran, Iraq; Libya, Algeria, Nigeria, Indonesia, Venezuela, and Ecuador. By 1970 OPEC had become the strongest entity in the world oil market, due to its large output (see Table 1-3) and the aggressive actions of its members' governments. Of the 16 bil.bbl. increase in oil output from 1952 to 1974, OPEC accounted for 10 bil.bbl.; 80 percent of the OPEC's share came from Middle East members.

A more detailed description of 1973 oil production and consumption is shown in Table 1-4.

The main characteristics of the oil market are:

1. OPEC members produced 55 percent of the world's output. Their exports constituted 96 percent of international oil trade.

Table 1-3
Main Sources of Oil Output, 1952-1974

Year	OPEC Middle East Bil. Bbl.	Percent	Total OPEC Bil. Bbl.	Percent	Other Bil. Bbl.	Percent	World Output Bil. Bbl.	Percent
1952	0.7	16	1.5	33	3.0	67	4.5	100
1960	1.9	25	3.2	42	4.5	58	7.7	100
1965	3.5	31	5.3	47	6.1	53	11.4	100
1970	6.2	35	8.6	49	9.0	51	17.6	100
1973	8.5	41	11.4	55	9.3	45	20.7	100
1974	8.3	42	11.1	56	8.9	44	20.0	100

Sources: 1952-1970: *anvario espanol del petroleo*, Anespe, 1971, pp. 240-241; 1973: *BP Statistical Review of the World Oil Industry, 1973*, London, 1974. 1974: *World Oil*, August, 1975.

Table 1-4
World Production and Consumption of Oil, 1973

	Production		Consumption		Net Export		Net Import	
	Bil. Bbl.	Percent	Bil. Bbl.	Percent	Bil. Bbl.	Percent	Bil. Bbl.	Percent
OPEC Middle East	8.5	41	0.5	2	8.0	71		
OPEC Other	2.9	14	0.1	1	2.8	25		
OPEC Total	11.4	55	0.6	3	10.8	96		
United States	3.8	18	5.9	29			2.1	19
Western Europe	0.2	1	5.5	26			5.3	47
Japan			2.0	10			2.0	18
Canada	0.7	4	0.6	3	0.1	1		
Communist Countries	3.5	17	3.2	15	0.3	3		
Developing Countries	0.9	4	2.3	11			1.4	12
Others	0.2	1	0.2	1				
Inventory Build-up			0.4	2			0.4	4
Total	20.7	100	20.7	100	11.2	100	11.2	100

Source: *B.P. Statistical Review of the World Oil Industry, 1973*, London, 1974.

2. Western Europe and Japan are pure consumers. Their imports accounted for 65 percent of international oil trade.
3. The United States, which is the largest consumer, produced only 63 percent of its own consumption. Its imports constituted 19 percent of world trade.
4. The developing countries imported 12 percent of world trade.
5. Canada, the communist countries, and the rest of the world are self-sufficient with insignificant surpluses.

The Main Participants

As we have seen, the main participants in the world oil market are the OPEC members, the United States, and the big pure-consuming countries.

OPEC Members

The OPEC members are divided into four groups:

1. Saudi Arabia, Kuwait, and the Persian Gulf Emirates (hereafter denoted as the S countries)
2. Other Arab countries (Libya and Iraq—the LQ countries)
3. Iran
4. Non-Middle East countries.

The first three groups together will be referred to as the Middle East OPEC members. Each country's relative power in the oil market depends on its share of total output and the size of its proven reserves. OPEC members possess about 70 percent of the world's proven oil reserves. Table 1-5 shows the distribution of OPEC reserves. The S countries hold 60 percent of OPEC reserves, while Iran possesses only 13 percent. The country with the largest reserves is Saudi Arabia with 165 bil.bbl.—one-third of OPEC reserves, and almost one-fourth of proven world reserves. These figures are official; unofficial data indicate that Saudia's reserves are much greater—as much as 300 to 450 bil.bbl. If the un-

Table 1-5
OPEC Oil Reserves, 1974

Country	Reserves (bil. bbl.)	Percent of OPEC	Percent of Middle East OPEC
Saudi Arabia	165	34	39
Kuwait	73	15	17
Abu Dabi	30	6	7
Other Gulf Emirates	32	6	7
Total S Group	300	61	70
Iran	66	13	16
Libya	27	6	6
Iraq	35	7	8
Total LQ	62	13	14
Total Middle East	428	87	100
Nigeria	21	4	
Algeria	8	2	
Indonesia	15	3	
Venezuela	15	3	
Ecuador	3	1	
Total Non-Middle East	62	13	
Total OPEC	490	100	

Source: *Oil and Gas Journal,* December, 1974.

official figures are true, this would have a far-reaching effect on the oil market (as will be seen below).

The S countries are by far the most important OPEC oil producers; in 1974 they provided 44 percent of OPEC output. Saudi Arabia is the largest single producer with 27 percent; all of the Arab countries together accounted for 55 percent. Iran is the second largest OPEC producer with 20 percent. All of the Middle East members together provided 75 percent of OPEC output (see Table 1-6).

As was seen in Table 1-4, OPEC members export 96 percent of the world's oil imports—which enabled OPEC to attain monopolistic power. In order to maintain such power, OPEC members must act together as a cartel. However, internal conflicts among the members limit OPEC's stability. It is perhaps more appropriate to describe OPEC as a price leader oligopoly in which major countries like Saudi Arabia and Iran compete for leadership. OPEC indeed operates as a cartel, but it is also prone to internal struggles.

Table 1-6
OPEC Oil Production, 1974

	Output (mil. bbl.)	Percent of OPEC	Percent of Middle East
Saudi Arabia	2,997	27	36
Kuwait	831	7	10
Abu Dhabi	515	5	6
Other Gulf Emirates	535	5	7
Total S	4,878	44	59
Iran	2,198	20	26
Libya	554	5	7
Iraq	675	6	8
Total LQ	1,229	11	15
Total Middle East	8,305	75	100
Algeria	361	3	
Nigeria	816	7	
Indonesia	502	4	
Venezuela	1,086	10	
Equador	64	1	
Total Non-Middle East	2,829	25	
Total OPEC	11,134	100	

Source: *World Oil*, August, 1975.

The United States

The United States consumes 29 percent of the world's oil output. It produces about 63 percent of its needs and imports 37 percent. In 1973 its imports constituted 9 percent of the world's imports—the highest of any country. Moreover, its imports have been increasing rapidly over the years because of the growing gap between its production and consumption. Due to the high level of consumption in the United States, a relatively modest increase in its consumption (while its output remained stable or even declined slightly) strongly affected the world market.

The United States has potential resources for increasing its oil output. The cost of producing from these sources ranges from $4/bbl. to $9/bbl. and more. There was no incentive to develop these sources when the market price was $3/bbl., but with market prices of $10/bbl. to $12/bbl., the incentive exists.

Western Europe and Japan

These countries are the main importers and consumers of OPEC oil. Although they are a small group, they have not utilized their oligopsonistic power to reduce oil prices. The Western European countries have been trying to develop their own oil sources in the North Sea in order to be less dependent on OPEC oil.

The Communist Countries

This book does not deal with the communist countries. They are practically self-sufficient in energy and oil, so their net oil trade with the western world is small. At higher prices these countries may become net exporters, and at lower prices they may become net importers—but only to a relatively insignificant degree.

Recent Developments

Until the 1970s the major oil corporations controlled most of the western world's production. As the market power of the OPEC governments increased, the impact of the big corporations diminished. This became particularly noticeable when the American oil companies were forced to participate (at least officially) in the 1973 oil embargo on their own country.

From 1970 to 1973 OPEC and the oil corporations were engaged in a series of negotiations and agreements through which oil prices were gradually increased. However, the governments of consuming countries did not take any precautions to meet the expected implications of these price rises. Moreover, the oil corporations planned to use more Middle East oil to supply—at low prices—the rapidly increasing consumption. Thus the 1973 oil embargo came as a shock to both the consuming countries and the oil corporations.

In October 1973 OPEC increased the posted price of oil to $5/bbl. In December OPEC announced another increase to $11.65/bbl. effective in January 1974. After a few months of uncertainty and price fluctuation at even higher levels, the actual oil prices stabilized in the $10.50/bbl. to $12.50/bbl. range. Certain official changes in September and November 1974, especially with regard to the royalties and tax rates paid to the OPEC governments, did not have a material effect on the actual oil prices. However, world inflation amounting to 20 to 25 percent from January 1974 to the end of 1975 has reduced the real price; the 1974 oil price in real terms would require an adjustment of the nominal price to $13/bbl. to $14/bbl. as of 1976. In September 1975 OPEC posted a 10 percent price increase that made up for only part of the accumulated inflation effect from January 1974.

The high prices affected the oil demand; world consumption in 1975 was not greater than that of 1973, while at low prices it could have been as much as 10 percent greater.

The outlook for the oil market is unclear. Lower consumption and greater competitive supply may indeed weaken OPEC's monopolistic power. But the effect of such possible development on the price level is not certain in view of potential internal conflicts among the major OPEC members on one hand and among the consuming countries on the other hand. The following chapter discusses a methodology for investigating future price developments in order to provide better insight into the problem.

Notes

1. *World Energy Supplies,* 1956-1959, series J, no. 4, United Nations, New York, 1961.

2. *B.P. Statistical Review of the World Oil Industry, 1973,* The British Petroleum Company, London, 1974.

3. *Der Energie Aubenhandel Westeuropaischer,* B.P. Benzin und Petroleum aktiengesellschaft Abteilung Volkswritschaft, London, 1969, p. 30.

2 Theoretical Background

Oil is a nonreproducible product belonging to the large family of exhaustible natural resources. Economic theory has not given appropriate attention to this group of products. The theory of production deals mainly with reproducible commodities; the theory of exhaustible resources is still in its embryonic stage.

The first attempt to develop a theory of nonreproducible resources was made by Professor Hotelling in his classic paper on exhaustible assets.[1] Published forty-five years ago, it remained a lone contender in the field until recently. Despite all the developments in economic theory and analytical tools since 1931, one may feel a sense of inferiority upon realizing how up-to-date this article really is. Little has been meaningfully added to Hotelling's comprehensive treatment.

With increasing awareness of the importance of exhaustible commodities—particularly our sources of energy—interest in the theory of exhaustible assets has also increased. Studies and articles have been published;[2] however, the theory has not yet been elaborated upon to a satisfactory degree.

This book attempts to apply dynamic methods of analyzing an oligopolistic market structure to the world energy problem. In this chapter we briefly review economic theory of exhaustible assets and then explain the methodology for our analysis of developments in oil prices.

Price Theory of Exhaustible Assets

The economic theory of exhaustible assets is dynamic—that is, it deals with the price determination of exhaustible assets over time. The following is a brief review of this theory in three different market structures: free competition, monopoly, and oligopoly.

Free Competition

Free competition in the market of an exhaustible asset is a state in which the asset is extracted by a great number of producers. Each producer can individually decide how much of the product he will offer for sale during each time period. Obviously, the more sold now, the less that will be left for sale in the future. If the market price is expected to remain constant, a producer would

11

prefer to sell as much as possible now so that maximum revenue could be taken in during the earliest period and then invested in secured interest-bearing financial assets (such as government bonds). This interest income gives a producer greater total profit on the earlier sales than on the later sales; the rate of interest on the interim investment determines how much is gained by selling earlier.

However, producers cannot increase total present sales and still maintain a constant price. As output and sales rise, prices will decline. This reduces the attraction of raising present sales to a maximum and makes it possible for the market to reach an equilibrium between output and price. In equilibrium, the producer will be indifferent to whether an additional unit is sold now or in the future; this occurs when the expected future price exceeds the present price by a rate equal to the interest rate. In this situation, selling a unit in the future instead of now has no effect on the total of the producer's present and future income; the interim interest is lost on the foregone present revenue but will be fully compensated by the greater revenue of increased future output at a higher price. Thus the equilibrium price in free compeition is not constant over time, but increases. The net price (price minus cost) rises at a rate equal to the interest rate. This pattern can be expressed in the following equation:

$$P_{t+1} = P_t + r(P_t - C_t) \qquad (2.1)$$

where P = unit market price
$\quad\ \ C$ = unit production cost
$\quad\ \ r$ = interest rate
$\quad\ \ t$ = year of sale
$(P_t - C_t)$ is clearly the net price in year t.

Monopoly

A monopoly is a market in which there is only one seller. The seller is able to determine either the price of the product or the quantity sold. Prices (or outputs) presumably are determined so that maximum profit over time is gained. Total profit is expressed in terms of its present value, i.e., the discounted value of all future profits.

The price pattern which provides the maximum present value of profit has certain mathematical characteristics. One is that marginal profit (the difference between marginal revenue and marginal cost) rises at a rate equal to the interest rate. As in free competition, the producer cannot increase the present value of profit by shifting a unit of output from the present to the future; the interim return that is lost on the foregone present marginal profit exactly offsets the greater marginal profit earned on a future sale.

Under monopoly the price level at any period is greater than the corresponding level under free competition. Hotelling has shown that the free competition equilibrium is consistent with maximization of public welfare; a monopoly is therefore nonoptimal for social welfare.

Oligopoly

The present theory of exhaustible assets deals mainly with free competition and monopoly. Within these frameworks the analytical solution of market equilibrium and the resulting price pattern over time is relatively straightforward. Nevertheless, the real market structure of exhaustible assets seldom falls within one of these categories; more often, as in the case of oil, it is an oligopolistic market. Unfortunately, the theory of oligopoly is still in an early stage of development.

An oligopoly is a market with a small number of producers, each of whom may affect the price. Consequently, each producer must consider the competitors' reactions when planning output and price policy. Since the pattern of the reaction is usually uncertain, the market outcome is uncertain. It is therefore impossible to provide a general solution for market equilibrium under oligopoly. The only analytical solutions are those based on specific assumptions of reaction functions. Thus an analysis of the oil market depends on one's conception of the main participants' reaction patterns. We use a model of oligopolistic behavior which seems to represent the world oil market practice reasonably well. This model is called *price leadership*. The conditions under which price leadership is expected to develop are: (1) very few producers of a quite homogenous product, and (2) one of these producers is much larger and no less efficient than the others. In this situation the dominant producer can determine the price and can correctly assume that the remaining producers will adjust their output to that price. The price leader then determines his own output as a monopoly facing the remaining net demand (i.e., total demand minus the competitors' supply).

A special case of oligopoly is the *cartel,* in which the producers of a homogenous product agree on a common price policy and locations of market and output. When the cartel includes all of the producers, it can develop a monopolistic price policy. If it includes the main producers but not all the producers, it may behave as a price-leader oligopoly.

A cartel is usually an unstable organization. Each member gains by agreeing to collaborate. After the cartel has been formed, however, a member can be even better off by breaking the agreement, increasing output, and taking advantage of the high cartel price. If a number of members do this, the cartel will break down.

The oil market is an oligopoly, and the Organization of Petroleum Exporting Countries (OPEC) constitutes a cartel of the major oil producing countries. This cartel exercises price leadership. This is the market framework of our analysis.

Methodology

The Problem

A price theory of exhaustible assets should help us to understand past price behavior and possible future price developments. However, the line between scientific forecasting and sheer prophetic guesswork is a vague and easily violated boundary.

The dynamic nature of the energy market involves great uncertainties. The uncertainty of supply is inherent in new explorations, new energy forms, unpredictable technological developments, costs of production and transportation, political factors, and competitive market structure. Although the future of energy demand is not as erratic as that of its supply, its changes are also uncertain. Thus even if supply and demand forecasts are based on available data and analyzed by appropriate statistical techniques, they are still subject to serious potential error. We have chosen to avoid this path; instead we concentrate on a normative approach.

The Approach

We start with the assertion that the oil market is an oligopoly in which OPEC functions as a price-leader cartel—with OPEC determining a price policy designed to achieve a maximum value of its goals. Given OPEC goals, what is its optimum price policy? The application of economic theory and techniques to this question provides a normative price prediction.

This means that if OPEC has monopolistic power, if the actual demand and competitive supply functions are close to those we employ, if OPEC acts as a goal maximizer, and if its goals are correctly described in our goal function, then OPEC is expected to design a price policy that is predictable by a proper optimization model.

The normative forecast described here seems to have a very limited scope. Nevertheless its scope can be conceptually extended in line with positive economics. We postulate that OPEC's actual price policy is determined "as if" OPEC behaves according to the above definitive statement. The resulting normative prediction then becomes a hypothesis to be tested against actual data in order to validate (or rather, invalidate) the "as if" theory.

The problem is further complicated by the fact that OPEC is not a coherent body. The normative predictions therefore may vary among the individual countries. The price policy actually determined by OPEC can thus be regarded as a compromise between optimal policies of conflicting subgroups.

Although the prediction of OPEC prices is a very complex problem, the normative approach provides valuable insight into the problem—leading to a better understanding of the complexity.

The Basic Model

A model that applies the normative approach to our problem is based on dynamic programming technique. The general framework of this application is as follows:

Demand. The world demand for energy depends on the level of real income and the price of energy. Thus one year's demand differs from that of the previous year by an amount that varies with the changes in real income and in the price level. A specific structure for the demand function and the dependence of the function on income and price changes will be described in Chapter 4. The general form of the function is as follows:

$$Q_t = F(Y_t, P_t) \qquad (2.2)$$

where Q_t = world energy demand in year t
 Y_t = world real national income in year t
 P_t = energy price in year t

Supply. The world supply of energy is composed of three sources:

 S^1 = supply of non-oil energy, i.e., coal, gas, nuclear energy, and synthetic fuel
 S^2 = supply of oil from all non-OPEC sources
 S^3 = OPEC oil supply

The non-oil supply S^1 varies over time with energy prices as follows:

$$S_t^1 = S_{t-1}^1 f^1(P_{t-1}) \qquad (2.3)$$

This states that the supply in year t depends on the supply in the previous year and the price in the previous year. $f^1(P_{t-1})$ specifies the pattern of the price effect.

The non-OPEC oil supply S^2 varies in a like manner:

$$S_t^2 = S_{t-1}^2 f^2(P_{t-1}) \qquad (2.4)$$

An estimate of these supply functions is presented in Chapter 3. The OPEC supply, S^3, is an endogenous factor of OPEC policy; thus it is incorporated in the model.

The Market for OPEC. Assuming a model of oligopoly price leadership, the demand for OPEC oil (hereafter referred to as D) is the net world demand for energy after deducting S^1 and S^2. *Therefore:*

$$D_t = Q_t - S_t^1 - S_t^2 \qquad (2.5)$$

Equation 2.5 is dynamic, since D_t, the demand in year t, depends on the energy prices in both year t (by means of Q_t) and year $t-1$ (by means of S_t^1 and S_t^2).

OPEC Price Policy. OPEC is assumed to act as a monopoly with respect to the demand for OPEC oil as expressed by Equation 2.5. Given a certain planning period, OPEC thus determines the price pattern over the period that maximizes its goal function. This goal function represents OPEC's net gain from its oil reserves—defined as the present value of two components:

1. Net revenues from oil sales (i.e., the difference between total revenue and cost) over the planning period.
2. The oil reserves remaining at the end of the planning period.

The solution of the model (i.e., the price pattern over time that maximizes the goal function) is referred to as the optimal price policy. This solution varies with assumptions about future demand and competitive supply as well as with assumptions about the goals and their relative importance. Since these assumptions are subject to uncertainty, the optimal solution is also subject to uncertainty.

In Chapter 6 we present specific optimal solutions of oil prices. Though one may be tempted to regard them as objective forecasts of oil prices, we would rather refer to them as preliminary exercises which serve as indicators of how normative prediction can be applied.

The same model can be applied not only to OPEC as a whole but also to individual OPEC members and subgroups (see Chapter 7). However, this may sometimes result in optimal price solutions which differ among the countries—creating potential economic conflict within OPEC that can also be linked with political factors. Thus the approach provides a basis for integrating economic and political factors as an explanation for past policy; it also indicates possible future policies for both the oil-exporting and oil-consuming countries.

Notes

1. Hotelling H., "The Economics of Exhaustible Assets," *Journal of Political Economy*, April 1931, pp. 137-175.

2. A few of these publications are: Adelman, M.A., *The World Petroleum Market*, Johns Hopkins University Press, Baltimore, 1972; Nordhaus, W.D., "The Allocation of Energy Resources," *Brookings Papers on Economic Activity*, 3, 1973, pp. 529-570; Sollow, R.M., "The Economics of Resources or the Resources of Economics," Richard T. Ely Lecture, *American Economic Review, Papers and Proceedings*, May 1974, pp. 1-14; Dasgupta, P., and Heal G., "The Optimal Depletion of Exhaustible Resources," *Review of Economic Studies, Symposium*, 1974, pp. 3-28; Levary D., and Liviatan N., "Notes on Hotelling's Economics of Exhaustible Resources," Falk Institute, (Jerusalem, 1975), Discussion Paper 751; Barnea A., and Lieber Z., "Dynamic Optimal Pricing to Deter Entry under Constrained Supply," mimeo, Tel-Aviv University, 1975.

3

Supply of Energy and Oil

Basic Concepts

The supply of energy is a function of its price. At low prices the amount supplied by producers is relatively small. As energy prices increase, there is greater incentive to invest in exploration and development in order to increase production; the higher prices cover the higher incremental costs plus profit. Thus the supply of energy is not a single specific quantity but a *schedule* of quantities that usually increase with price.

The supply of energy is the total of the supply of all energy forms—including oil, gas, coal, water power (hydroelectricity), nuclear energy, and solar energy. The supply of oil is the sum of the supply of oil from oil fields and the supply of synthetic fuels from shale oil, tar sand, and the liquefaction of coal. The extraction of synthetic fuels requires certain production processes which add to their cost; therefore these forms enter the supply schedule only at high prices.

Since the development of new oil fields and synthetic fuels takes time, it is worthwhile to distinguish between short-term supply and long-term supply. The short-term supply shows the immediate relationship between output and price before any new development takes place. The long-term supply includes additional sources that can be developed over time as a result of price changes.

This reasoning can be presented in a more concrete form by introducing two concepts of oil supply—*static* and *dynamic*. The static concept shows the relationship between the price and the amount supplied during a given period. The dynamic concept shows the relationship between the price level and the *rate of change* in the supply schedule over time.

The effect of price on the amount supplied is measured by the price elasticity of supply. *Elasticity* is a coefficient that shows the percentage increase in the amount supplied resulting from a 1 percent price rise (or conversely, the percentage by which supply is reduced as a result of a 1 percent price decline).[a]

---

[a]The price elasticity of supply (n_s) is measured by the following formula:

$$n_s = \frac{\Delta Q/Q}{\Delta P/P}$$

where Q is the quantity supplied, P is the price level, and ΔQ is the change in quantity resulting from a ΔP change in price level. This formula can be rewritten as:

$$n_s = \frac{\Delta Q}{\Delta P} \cdot \frac{P}{Q}$$

19

Elasticity is applicable to both the static concept of supply (in which case we refer to short-term elasticity) and the dynamic concept (long-term elasticity). These two concepts, illustrated in Figure 3–1, apply to energy as a whole as well as to any specific form of energy.

The static version (A) is a schematic representation of supply as a function of price at a given time. The dynamic version (B) is a schematic representation of the rate of increase in output over time (from period 0 to period T) for different price levels. This particular scheme assumes that each price level will be constant during the given time. If we assume nonconstant price patterns over time, the rates of output increase will vary—and the shapes of the curves will change.

Estimates of the relationship between future energy prices and expected rates-of-output increase are subject to great uncertainty. In making these indicative estimates we have used the following procedure: First the supply of each energy source for 1980 (and 1985 when possible) is estimated on the basis of available information at alternative price levels held constant from 1974 to 1980. Then the estimates are used to infer the average *annual* (long-term) rate of change in supply at each price level. This procedure is applied to non-oil energy sources and to non-OPEC oil sources.

A unique situation exists regarding OPEC oil in the Middle East. Tremendous oil reserves were found with a relatively low investment. The cost of extracting the oil has been only \$0.10/bbl. to \$0.20/bbl., including recovery of investment. New oil fields may also be discovered that will increase total proven reserves in the long run. Thus it is possible that the Middle East's

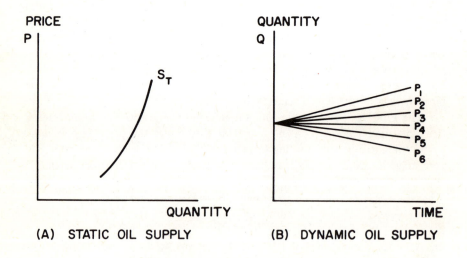

Figure 3-1. Schematic Oil Supply

maximum output of oil at very low prices is much greater than the present level.

This is the background for two totally different market scenarios. The first is free competition, in which the Middle East countries and the other OPEC members compete among themselves and with the rest of the oil-producing countries. They offer their maximum output at a price equal to their marginal cost of production, and the world market has plenty of cheap oil for a long period until the Middle East reserves are depleted.

The second scenario, which more properly describes the present market structure, shows the Middle East countries as members of a price leader cartel (OPEC). The cartel sets its output so as to reach a price level that maximizes oil profits; thus the concept of a supply schedule does not apply to this cartel.

The subsequent analysis of the long-term energy supply follows the second market scenario; it excludes the OPEC members. OPEC output will be determined in the context of its policy decisions, which are analyzed in Chapters 6 and 7.

Energy Supplies for 1980 and 1985

The estimated energy supplies for 1980 and 1985 are broken down into oil and non-oil sources. The main sources of the non-OPEC oil supply are United States, "Normal 48" (the first 48 states); United States, new sources; Western Europe; the rest of the world (including Canada, Latin America, and others, but excluding the communist countries—see Chapter 1); and the synthetic fuels. The non-oil energy sources are mainly coal and natural gas—plus other sources like hydroelectricity, nuclear energy, and solar energy.

Oil

U.S. Normal 48. Total output in 1972 was 3.4 bil.bbl. continuing a decline of about 2 percent per year. (Output has been decreasing due to the exhaustion of reserves.) The 1972 market price was under $3/bbl. Studies in the 1960s[1] indicate that the price elasticity of supply from this source is between 0.3 and 0.8. This is the sum of the price elasticities of wildcat drillings, success ratio, and size of discoveries. These elasticity estimates refer only to the lower price range. Therefore our estimates for 1980 assume that at a price of $3/bbl., output will continue to decline by 2 percent per year. For higher price levels, the 1980 output is estimated by applying the lower coefficient of long-term supply elasticity (0.3) to the $3/bbl. output level. Specifically, the amount that would be supplied in 1980 at $3/bbl. is estimated at 87 percent of the 1973 supply—2.6 bil.bbl. The supply at higher prices is estimated in Table 3-1.[b]

These estimates might be too high. For example, the Federal Energy Administration (FEA) estimates of Normal 48 output are considerably lower. It estimates the 1980 output at 2.1, 2.2, and 2.4 bil.bbl. at prices of $4/bbl., $7/bbl., and $11/bbl. respectively. (At the same prices 1985 output is estimated at 2.1, 2.6, and 3.3 bil.bbl. respectively.) The FEA's estimated price elasticity of supply for 1980 is extremely low: .04 to .20.

U.S., New Sources.[2] The new sources are:

Secondary and Tertiary Recovery of Old Oil Fields. The normal technique extracts only about one-third of the oil in the ground. The remaining recoverable reserves are estimated at 30 bil.bbl. Existing techniques for further recovery will yield oil at estimated cost of $10/bbl. However, assuming that by 1985 more efficient techniques will be available, it may be possible to extract about half of the remaining reserves at a cost of $5/bbl. to $6/bbl. Output in 1985 is thus estimated at 0.5 bil.bbl.—with another 0.5 bil.bbl. possible at $10/bbl. No output from this source is estimated for 1980.

Offshore. The reserves are estimated at between 10 bil.bbl. and 50 bil.bbl. Output is estimated 0.5 bil.bbl. in 1980 and 1 bil.bbl. in 1985 at a cost of $4/bbl. to $5/bbl.

Alaska–Prudhoe Bay. Reserves are estimated at 15 bil.bbl. to 30 bil.bbl. At a price of $4/bbl. to $5/bbl., 1980 output is estimated at 1 bil.bbl. (The FEA's unofficial estimates are 3 bil.bbl. in 1980 and 5 bil.bbl. in 1985.)

Table 3-1
Estimated 1980 Supply Schedule of Normal 48

Price $/bbl.	Output (bil. bbl.)
3	2.5
4	2.7
5	3.0
6	3.2
7	3.3
8	3.5
9	3.6

---

[b]This estimate is based on the following arc-elasticity formula, which approximates the measure of elasticity when prices change over a wide range:

$$n_s = \frac{Q_2 - Q_1}{P_2 - P_1} \cdot \frac{P_1 + P_2}{Q_1 + Q_2}$$

Alaska–Naval Reserves. A rough exploration indicates the probability of tremendously high reserves of 25 bil.bbl. to 120 bil.bbl. It is assumed that output from this source will not begin before the 1980s. In 1985 output may reach 1 bil.bbl. or 2 bil.bbl. at cost of $5/bbl. to $6/bbl.

Out-Continental Shelf. This source has not yet been explored. It is assumed that recoverable reserves will be about 30 bil.bbl. in 1985 and that the cost will be $5/bbl. to $6/bbl. No output is expected in 1980.

Other Sources. It is reasonable to assume that at high prices the incentive to explore further sources of oil will result in some unexpected recoveries. We therefore "guesstimate" that by 1985, 1 bil.bbl. of annual output will appear at a price range of $6/bbl. to $8/bbl. However, no output from such sources is expected in 1980.

This brief review shows that no significant increase in U.S. oil output from new sources should be expected by 1980. At high prices, however, output will increase substantially after 1980. The expected outputs in 1980 and 1985 are shown in Table 3-2.

Table 3-2
Estimated New Sources of Oil in the United States: 1980, 1985
(In Billions of Barrels)

Price ($/bbl.)	Source	Output 1980	Output 1985	Reserves 1985
4-5	Offshore	0.5	1.0	10-50
	Alaska-Prudhoe Bay	1.0	1.0	10-30
	Alaska-Naval Reserve	—	1.0	25-120
	Total	1.5	3.0	45-200
5-6	Secondary and Tertiary Recovery		0.5	15
	Out-Continental Shelf		1.0	30
	Total		1.5	45
6-8	Tertiary Recovery		0.5	15
	Other Unspecified Sources		1.0	30
	Total		1.5	45

Total U.S. Supply. Total supply of the U.S. in 1980 and 1985—the sum of the Normal 48 supply plus new sources—is shown in Table 3-3.

This long-term supply schedule is simply a reference base. It is subject to many qualifications, since it is based on a large number of assumptions and "guesstimates." Indeed it differs greatly from the FEA estimates. We feel, however, that the FEA projections underestimate the effectiveness of the high-price incentive on the long-term development of oil output.

Western Europe. Information on which to base estimates for Western Europe is extremely limited. Its main source of oil is the North Sea. Houthakker and Kennedy[3] assume that by 1980 the annual oil output of the North Sea will be nearly 1.5 bil.bbl. Odell[4] indicates that Western Europe's output will be 2.2 bil.bbl. in 1980 and 2.75 bil.bbl. in 1985. Both assume a price of at least $5/bbl. These references are the basis for the rough estimates in Table 3-4.

The implicit assumptions here are:

1. At low prices there will be some development in the 1970s. If prices remain low, the development will stop in the 1980s.
2. For technological reasons it is difficult to increase output beyond 1.5 bil. bbl. by 1980. By 1985, however, it will be possible to utilize the capacity predicted by Odell.

Rest of the World. The 1973 oil output of the rest of the world was 2.6 bil.bbl.[c] We assume that at a low price ($3/bbl.) 1980 output will be somewhat lower (2.4 bil.bbl.); at a higher price ($8/bbl.) 1980 output will be greater (3.0 bil.-bbl.)

Table 3-3
Estimated Total U.S. Supply Schedule: 1980, 1985
(Output in Billions of Barrels)[a]

Price ($/bbl.)	1980 Output	1985 Output[b]
3	2.5	2.5
4	3.0	3.0
5	4.5	6.0
6	4.7	7.5
7	4.8	8.5
8	5.0	9.5

[a]The figures are rounded in a direction that "smooths" the shape of the schedule.

[b]We assume that the 1985 supply of Normal 48 will be the same as in 1980.

[c]The communist countries are excluded (see Chapter 1).

Table 3-4
Estimated Oil Output in Western Europe: 1980, 1985
(In Billions of Barrels)

Price ($/bbl.)	1980 Output	1985 Output
3	0.5	0.5
4	1.0	1.0
5-8	1.5	2.75

Synthetic Fuels. Technologically, it is possible to produce oil from various sources at an estimated cost of $8/bbl. to $10/bbl. At this cost there is no long-term economic constraint on output to fill the gap between conventional oil sources and world demand. The cost estimate may change with the gaining of practical experience.

Development of these sources will take time. We assume that at a price above $10/bbl. unlimited output can be reached between 1985 and 1990. This assumption is based on technological and economic feasibility—disregarding political and institutional barriers. If development programs are not pushed vigorously, the 1985-1990 supply may not be unlimited. In 1980 the output of oil from substitute sources will be limited. We assume that it might reach a level of 0.5 bil.bbl., all produced in the U.S.

Non-Oil Energy

Coal. Coal output in 1973 (excluding the communist countries) amounted to 1,560 million tons, which is equivalent to about 6.75 bil.bbl. of oil. The share of coal declined from 62 percent of total energy consumption in 1950 to 30 percent in 1973. However, at an equivalent price of $4/bbl. the share of coal is expected to stop declining, and its output is expected to increase at an average annual rate of about 2 percent. Following Walter Levy's estimates,[5] we assume that at an equivalent price of $7/bbl. coal output will rise 4 percent per year; at higher prices output will probably rise at a greater rate. The estimated 1980 coal supply is shown in Table 3-5.

Natural Gas. Output of natural gas in 1973 (excluding communist countries) was 34 trillion cubic feet (the equivalent of 6 bil.bbl. of oil). We assume that the annual rate of increase will range from 2 percent at an equivalent price of $4/bbl. to 5 percent at an equivalent price of $7/bbl. to $8/bbl. Thus we have the indicative supply schedule for 1980 (Table 3-6).

Other Sources of Energy. Other sources of non-oil energy—such as hydroelectricity, nuclear energy, solar energy, etc.—amounted to 2.5 percent of the total

Table 3–5
Estimated Coal Supply: 1980

Price ($/bbl.)	Quantity (bil.bbl. equiv.)
3	6.75
4	7.75
5–7	9.00

Table 3–6
Supply Schedule of Natural Gas: 1980

Price ($/bbl.)	Quantity (bil.bbl. equiv.)
3	6.0
4	6.9
5	7.4
6	7.9
7–8	8.5

1973 output of energy (the equivalent of about 0.8 bil.bbl. of oil). We assume that in 1980 these sources will reach the equivalent of 1.2 bil.bbl.

Time Pattern of Energy Supply

On the basis of the supply estimates for 1980 and 1985 for various price levels, it is now possible to infer the effect of energy prices on energy supply over time.

Non-Oil Energy Supply

The total 1973 non-oil energy supply was the equivalent of 13.6 bil.bbl. of oil. The estimated annual rates of change at alternative price levels are shown in Table 3–7. At higher prices the rate of change continues to rise linearly. The functional relationship is as follows:

$$a_t = \begin{cases} -0.06 + 0.020P_{t-1} & \text{for } P_t \leqslant \$4 \\ 0.005P_{t-1} & \text{for } P_t \geqslant \$4 \end{cases} \tag{3.1}$$

where a_t indicates the rate of increase at year t.

Table 3-7
Estimated Annual Rates of Change of Non-Oil Energy Supplies

Price ($/bbl.)	Annual Rate of Change in Supply
3	0.000
4	0.020
5	0.025
6	0.030

The non-oil energy output of year $t [S_t^1]$ is a function of the previous year's output and price:

$$S_t^1 = (1 + a_t)S_{t-1}^1 \qquad (3.2)$$

Accordingly we have:

$$S_t^1 = \begin{cases} (0.940 + 0.020P_{t-1})S_{t-1}^1 & \text{for } P_t \leqslant \$4 \\ \\ (1.000 + 0.005P_{t-1})S_{t-1}^1 & \text{for } P_t \geqslant \$4 \end{cases} \qquad (3.3)$$

Non-OPEC Oil Supply

The total 1973 output of non-OPEC oil was 5.7 bil.bbl. The estimated annual rates of change in supply at alternative price levels are shown in Table 3-8.

At higher prices a linear relationship is assumed to persist. This is approximated by the following function:

$$b_t = \begin{cases} -0.101 + 0.027P_{t-1} & \text{for } P_t \leqslant \$5 \\ \\ 0.010 + 0.005P_{t-1} & \text{for } P_t > \$5 \end{cases} \qquad (3.4)$$

where b_t indicates the rate of increase in supply at year t.

Table 3–8
Estimated Annual Rates of Change of Non-OPEC Oil Supplies

Price ($/bbl.)	Annual Rate of Change in Supply
3	-0.020
4	0.007
5	0.034
6	0.040
7	0.044

Thus the non-OPEC oil output in any year $t[S_t^2]$ is given by:

$$S_t^2 = (1 + b_t)S_{t-1}^2 \tag{3.5}$$

and it clearly depends on the previous year's output and price. Accordingly, we have:

$$S_t^2 = \begin{cases} (0.899 + 0.027P_{t-1})S_{t-1}^2 & \text{for } P_t \leqslant 5 \\ \\ (1.010 + 0.005P_{t-1})S_{t-1}^2 & \text{for } P_t > 5 \end{cases} \tag{3.6}$$

Equations 3.3 and 3.6 will be inserted into the dynamic model presented in Chapter 5.

Non-OPEC Energy Supply in 1980 and 1985

Estimates of energy supply (excluding the communist countries and OPEC) for 1980 and 1985 are presented in Table 3–9. These estimates are based on various prices assumed to remain constant after 1976. The projection uses the actual 1973 prices and applies $11/bbl. for 1974 and 1975. The data are divided into non-oil energy and non-OPEC oil. OPEC oil is the difference between expected demand and the supply presented in Table 3–9.

Given the 1973 non-oil output at the equivalent of 13.6 bil.bbl. of oil, output is expected to increase 20 percent by 1980 at low prices ($4/bbl.) and 40 percent at high prices ($12/bbl.). The 1985 increases are estimated at 33 percent and 90 percent respectively.

Non-OPEC oil output in 1980 is expected to exceed that of 1973 by only 14 percent at low prices ($4/bbl.) but by 45 percent at high prices ($12/bbl.). By 1985 the increase will amount to only 17 percent at low prices but will more than double at high prices.

Table 3-9
Estimated Non-OPEC Energy Supply: 1980, 1985
(Bil. Bbl. Oil Equivalents)

	1980		1985	
Price ($/bbl.)	Non-Oil	Non-OPEC Oil	Non-Oil	Non-OPEC Oil
4	16.4	6.5	18.1	6.7
5	16.7	7.3	18.9	8.6
6	17.0	7.4	19.8	9.0
7	17.4	7.6	20.6	9.4
8	17.7	7.7	21.5	9.8
9	18.1	7.9	22.5	10.3
10	18.4	8.0	23.5	10.7
11	18.8	8.2	24.5	11.2
12	19.1	8.3	25.6	11.7

Notes

1. Fisher, F.M., *Supply and Costs in the U.S. Petroleum Industry, Two Econometric Studies,* Johns Hopkins Press, Baltimore, 1964; Erickson, E.W., and Spann R.M., "Price, Regulation and the Supply of Natural Gas in the U.S., in *Resources for the Future,* Keith Brown, ed.; Mancke, R.M., "The Long-Run Supply Curve of Crude Oil Produced in the U.S.", *Antitrust Bulletin,* Winter 1970, pp. 727-56.

2. This section draws to a great extent from discussions with Herman Kahn.

3. Houthakker, H.S., and Kennedy, M., "Demand for Energy as a Function of Price," Mimeo 1973.

4. Odell, P.R., "The Availability of Indigenous Energy in Western Europe 1973-1998 with Special Reference to Oil and Natural Gas," 1st World Symposium, Energy and Raw Materials, Paris, June, 1974.

5. Walter Levy, "World Oil Cooperation or International Chaos", *Foreign Affairs,* July 1974.

Appendix 3A

Estimated 1980 Energy Supply
in the U.S.

Estimates of the 1980 energy supply for the U.S., prepared as part of the estimate of total world energy supply, are presented in Table 3A-1. The purpose of this appendix is to compare these estimates with those of an M.I.T. Study.[1]

The M.I.T. Judgmental Oil Supply Forecast for 1980 is 5 bil.bbl. at any price from $7/bbl. to $11/bbl. This forecast is very close to our estimate of 5 bil.bbl. at a price of $8/bbl. The M.I.T. econometric model for oil supply indicates somewhat lower estimates: 3.8 bil.bbl. at $7/bbl. and 4.6 bil.bb. at $9/bbl.

The M.I.T. estimates of total U.S. energy supply at a price of $7/bbl. (based on both the econometric model and the judgmental forecast) is 14.0 bil.bbl. Our respective estimate is 13.5 bil.bbl. The two M.I.T. estimates at $9/bbl. are greater: 14.3 bil.bbl. and 15.5 bil.bbl. Our estimate is 14.6 bil.bbl.

Our estimates of the U.S. energy supply in 1980 are applied (in Appendix 4A) to the U.S. demand forecasts to evaluate the circumstances under which the U.S. may achieve energy independence.

Table 3A-1.
Estimated U.S. Energy Supply, 1980
(Bil. Bbl. or Equivalent)

Price ($/bbl.)	Oil	Non-Oil Energy	Total
3	2.5	7.2	9.7
4	3.0	8.2	11.2
5	4.5	8.3	12.8
6	4.7	8.5	13.2
7	4.8	8.7	13.5
8	5.0	8.9	13.9
9	5.5	9.1	14.6

Note

1. Adelman M., et al, "Energy Self-Sufficiency, An Economic Evaluation," *Technology Review,* May 1974, pp. 23-52.

4

Demand for Energy and Oil

This chapter analyzes the factors affecting energy demand over time, and estimates future demand for energy and for oil. Similar estimates of U.S. energy and oil demand are summarized in the appendix to this chapter.

Factors Affecting the Demand for Energy

The demand for energy is affected mainly by two factors—the price of energy and the level of income.

The Effect of Price

Price is the main factor in the demand for any product; the lower the price, the greater the amount demanded. The price effect is measured by the price elasticity of demand. This coefficient shows the percentage by which the amount demanded rises as a result of a 1 percent price reduction (or conversely, the percentage by which demand is reduced as a result of a 1 percent price increase).[a] Since price and demand move in opposite directions, price elasticity is a coefficient less than zero.

---

[a] Let n_p = price elasticity of demand. The mathematical definition of n_p is:

$$n_p = \frac{\Delta Q/Q}{\Delta P/P}$$

where P is the price level, Q is the quantity demanded and ΔQ is the change in quantity resulting from a ΔP change in price. This formula can be rewritten as:

$$n_p = \frac{\Delta Q}{\Delta P} \cdot \frac{P}{Q}$$

When we have two points (1 and 2) on the demand function, the demand elasticity between the points is calculated by the following formula:

$$n_p = \frac{Q_2 - Q_1}{P_2 - P_1} \cdot \frac{P_1}{Q_1}$$

When the two points are not close enough to each other, the measure should be approximated by the arc elasticity formula as follows:

$$n_p = \frac{Q_2 - Q_1}{P_2 - P_1} \cdot \frac{P_1 + P_2}{Q_1 + Q_2}$$

After World War II the price of oil was low; it declined until 1967, especially in real terms (see Chapter 1). This price decline accelerated the increase in oil consumption by encouraging the use of oil and oil-intensive products; by speeding up the substitution of oil for coal (see Chapter 1); and by accelerating the development of energy-intensive production technologies and discouraging the development of energy-conserving devices.

The Effect of Income

As income rises, the demand for energy increases. The effect of rising income on demand is measured by the income elasticity of demand. This coefficient indicates the percentage by which demand rises as a result of a 1 percent rise in income.[b] The increase in national income in the Western countries after World War II obviously contributed to the increase in oil demand.

Empirical Estimates of Demand Elasticities

Past data on energy consumption are associated with a simultaneous decline in real prices and an increase in income. Therefore an analysis of consumption based only on income will overestimate the income effect because it incorporates the positive effect of declining prices on consumption. In order to separate the two effects, the two elasticities should be estimated simultaneously by one statistical method.

A forecast by the U.S. Federal Energy Office[1] indicates 1980 consumption levels for energy and oil at $4/bbl., $7/bbl., and $11/bbl. The inferred price elasticity of demand for the energy estimates is -0.10; for the oil estimates, -0.20.

Comprehensive estimates of price elasticity and income elasticity for oil were made by Houthakker and Kennedy.[2] The econometric model they apply provides estimates of short-term elasticity as well as a parameter that reflects the adjustment of demand over time. The long-term elasticity is derived from these two estimates. The model is applied to various energy products of different regions and countries in different time periods. A selected summary of the estimates is shown in Table 4-1.

---

[b]The mathematical definition of income elasticity of demand is:

$$n_y = \frac{\Delta Q/Q}{\Delta Y/Y} = \frac{\Delta Q}{\Delta Y} \cdot \frac{Y}{Q}$$

where Y is the income level.

Table 4-1
Estimated Long-term Demand Elasticities

	Price Elasticity	Income Elasticity
U.S. Gasoline	−0.24	0.98
OECD[a] Gasoline	−0.82	1.30
U.S. Residential Electricity	−1.00	1.60
OECD Residential Oil	−1.58	1.60

[a]Organization for Economic Cooperation and Development

Source: Houthakker, H.S. and Kennedy, M., "Demand for Energy as a Function of Price," Mimeo., 1973.

Houthakker and Kennedy imply price elasticities of aggregate oil demand on a country basis as follows:

Japan:	-0.75
Europe:	-0.20 to -0.40
U.S.:	-0.30

The price elasticity estimates for specific products are extremely high. The implied price elasticities for aggregate country demand are lower.

Houthakker's findings indicate a surprisingly high long-term price elasticity of oil demand. Even the lower range figure (about -0.30) has a substantial effect on future demand. If the price level in 1980 is three times higher than in 1973, oil demand in 1980 (for a constant income) will be about 25 percent lower than it would have been at lower constant prices. This also means that if the price of energy in 1970 had been at the real 1948 level (i.e., about double its actual 1970 level), consumption in 1970 would have been 20 percent lower than it actually was.

Many other studies also provide estimates of price and income elasticities for energy and oil.[3] These studies were made for different purposes at different time periods in different markets and are inconsistent. The elasticities vary between very low estimates (almost 0) to very high ones (significantly over 1). Thus, one should apply these findings with caution.

Moreover, the available findings are all based on studies made when energy and oil prices fluctuated within a *low* price range. Elasticity is not necessarily constant at different price levels; consequently, it is difficult to project elasticities from the past (when prices were low) to the future (when prices may be high).

In the following empirical analysis of *energy* demand we use a price elasticity of -0.10 and an income elasticity of 1.0 for our central estimates.[c] The demand for *oil* is derived from the combination of energy demand and non-oil energy supply. The elasticity of oil demand is greater than that of energy demand.

The Energy Demand Function

After examining statistical data and adjusting alternative demand functions, we selected a family of demand functions for energy that is schematically represented by Figure 4-1. The form of the function is as follows:

$$Q = Q_M + K \cdot P^{-n} \tag{4.1}$$

The specific shape of the function depends on its parameters: Q_M = a minimum essential energy demand; n = the price elasticity in the section to the *right* of Q_M (i.e., with regard to $Q - Q_M$); and K. While the price elasticity of this function with regard to $Q - Q_M$ is $-n$, the price elasticity with regard to Q

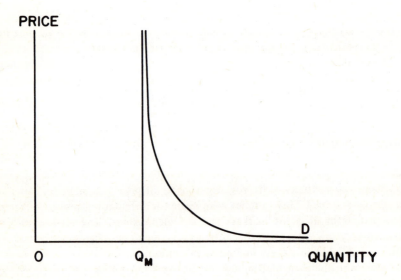

Figure 4-1. Schematic Energy Demand

---

[c]A sensitivity analysis of these assumptions appears below.

is $-n(Q - Q_M)/Q$, thus:

$$n_p = -n(1 - \frac{Q_M}{Q})$$ (4.2)

In order to adjust specific parameters to the demand function, one must consider historical data. Before October 1973, energy prices were low—under \$3/bbl. of oil equivalent. The price-elasticity estimates in the previous section apply to the demand function only in this price range. We employ a price elasticity n_p = -0.10 at the point of the function where Q_{73} represents the actual consumption level of 1973 (30.7 bil.bbl.) of the 1973 price level (P_{73}) of \$3/bbl. On the basis of the two equations (4.1 and 4.2) we thus establish a functional relationship between the three parameters K, Q_M and n.

Thus for each n it is possible to solve the values of Q_M and K, as shown in the following examples:

$$
\begin{aligned}
n &= 1.0 & Q_{73} &= 27.63 + 9.21p^{-1} \\
n &= 0.8 & Q_{73} &= 26.86 + 9.24p^{-0.8} \\
n &= 0.6 & Q_{73} &= 25.58 + 9.89p^{-0.6} \\
n &= 0.5 & Q_{73} &= 24.56 + 10.63p^{-0.5} \\
n &= 0.3 & Q_{73} &= 20.47 + 14.23p^{-0.3}
\end{aligned}
$$ (4.3)

This family of functions will be inserted into the dynamic model for forecasting the OPEC members' optimal price policy (see Chapter 5). In addition, it allows the estimation of energy prices for future periods (e.g., 1980) and their comparison with other forecasts.

Estimated 1980 Demand for Energy

Energy prices affect energy demand gradually. The effect is cumulative. Therefore short-term price elasticity is smaller than long-term elasticity. In order to demonstrate the estimated long-term effect of higher energy prices on demand, we assume that most of the price effect accumulates during the first seven years. This assumption is consistent with Houthakker's findings.

1980 demand is derived on the basis of Equation 4.1. The values of para meters Q_M and K are increased by a certain proportion. This proportion is the product of the rate of income increase and the income elasticity.

Assume, for example, an income rise of 4 percent per year. This will accumulate to 31.6 percent by 1980. Let n_y equal income elasticity; then:

$$Q_{1980} = (1 + 0.316 n_y)Q_M + (1 + 0.316 n_y)K \cdot P^{-n}$$ (4.4)

where Q_M and K are the parameters of the 1973 demand function in Equation 4.1.

The 1980 demand for energy depends not only on 1980 prices but also on the price pattern prior to 1980. Since the number of reasonable patterns is exceedingly large, such demand estimates can become tedious. We have chosen, however, to give an indication of such estimates—by assuming that for each price in 1980 the same price (in real terms) existed during the entire preceding period.

We first tested the sensitivity of the 1980 energy demand to changes of n for our central estimation of price elasticity (n_p = -0.10) and income elasticity (n_y = 1.0). The resulting energy demanded in 1980 at a price of \$7/bbl. varies between 38.1 bil.bll. (for n = 1.0) and 37.4 bil.bbl. (for n = 0.3). At a price of \$10/bbl. the respective range is 37.5 bil.bbl. and \$36.3 bil.bbl. We thus conclude that the demand for energy is quite insensitive to changes of n in the range between 1.0 and 0.3

In order to test the demand sensitivity to the price and income elasticity we selected a function where n = 1. The 1980 energy demand is found to be more sensitive to these latter parameters. This is presented in Table 4-2, which brings six demand estimates for 1980 for price elasticites of -0.10, -0.20, and income elasticities of 0.75, 1.0 and 1.25.

Table 4-2
Estimated World Energy Demand, 1980
(Bil. Bbl.)

Income Elasticity	0.75		1.00		1.25	
Price Elasticity	-0.10	-0.20	-0.10	-0.20	-0.10	-0.20
Price ($/bbl.)						
3	38.0	38.0	40.4	40.4	42.8	42.8
4	37.0	36.1	39.4	38.4	41.8	40.7
5	36.5	34.9	38.8	37.2	41.1	39.4
6	36.1	34.2	38.4	36.4	40.7	38.5
7	35.8	33.6	38.1	35.8	40.4	37.9
8	35.6	33.2	37.9	35.4	40.1	37.5
9	35.4	32.9	37.7	35.0	40.0	37.1
10	35.3	32.7	37.6	34.7	39.8	36.8
11	35.2	32.5	37.5	34.5	39.7	36.6
12	35.1	32.3	37.4	34.3	39.6	36.4
High	34.2	30.4	36.4	32.3	38.5	34.3

Total energy consumption in 1973 was 30.7 bil.bbl. Table 4-2 shows that at $11/bbl. for example, total demand for energy in 1980 is expected to exceed that of 1973 by 5.9 percent at the lowest estimate, by 29.3 percent at the highest estimate, and by 22.1 percent at the central estimate—equivalent to an annual 2.9 percent increase.

Estimated 1980 Demand for Oil

Demand for oil in 1980 is the difference between the 1980 demand for energy, estimated in the preceding section, and the 1980 supply of non-oil energy, estimated in Chapter 3. As with demand, the non-oil energy supply for 1980 also depends on the price pattern during this period. Therefore the assumption of constant prices made in the previous section also applies to the following examples.

Table 4-3 presents six alternative oil demand schedules for 1980. The schedules are consistent with the above six schedules of energy demand presented in Table 4-2. The estimates may be compared with the 1973 oil consumption of 17.1 bil.bbl.

These schedules show that the effect of price on oil demand is considerably greater than the effect of price on demand for energy as a whole. If the oil price remains at $11/bbl. in real terms and real income remains constant, total demand for oil in the central estimate is expected to be 29 percent lower than

Table 4-3
Estimated World Oil Demand, 1980
(Bil. Bbl.)

Income Elasticity	0.75		1.00		1.25	
Price Elasticity	−0.10	−0.20	−0.10	−0.20	−0.10	−0.20
Price ($/bbl.)						
3	24.0	24.0	26.4	26.4	28.8	28.8
4	21.7	20.8	24.1	23.1	26.5	25.4
5	20.7	19.1	23.0	21.4	25.3	23.6
6	19.9	18.0	22.2	20.2	24.5	22.3
7	19.1	16.9	21.4	19.1	23.7	21.2
8	18.4	16.0	20.7	18.2	22.9	20.3
9	17.7	15.2	20.0	17.3	22.3	19.4
10	17.1	14.5	19.4	16.5	21.6	18.6
11	16.4	13.7	18.7	15.7	20.9	17.8
12	15.8	13.0	18.1	15.0	20.3	17.1

at a price of $3/bbl. Due to the positive income effect, however, this demand will be greater than the 1973 consumption by 9.4 percent—the equivalent of a 1.3 percent annual increase in oil consumption.

Consequently, the price elasticity of demand for oil is significantly greater than that for energy as a whole. For example, for the central estimate at $P =$ $3/bbl., where the energy price elasticity is -0.10, the respective elasticity of oil demand at higher price levels varies between -0.20 and -0.35. This range of price elasticity turns out to be consistent with the lower estimate of the empirical fundings presented earlier in this chapter.

Notes

1. U.S. Federal Energy Office, "Fossil Fuel and Electricity Demand Forecast by Major Consuming Sectors: Basic Results and Summary Description," June 3, 1974.

2. Houthakker, H.S., and Kennedy M., *op.cit.* See also Houthakker, H.S., "The Price Elasticity of Energy Demand," mimeo, Committee for Economic Development, December 1974.

3. For a comprehensive summary of such studies, see Kuenne et al., "Intermediate Term Energy Programs to Protect against Crude-Petroleum Import Interruptions," Institute for Defense Analyses, Paper P. 1063, Sept. 1974, Table I.

Appendix 4A

Estimated 1980 Energy and Oil Demand in the U.S.

This appendix presents estimates of the U.S. demand for energy and oil based on the methods applied to the world as a whole.[a]

In 1973 U.S. consumption of energy amounted to 13 bil.bbl. (oil equivalent). Of this some 6 bil.bbl. were oil; the remaining 7 bil.bbl. were mainly coal and gas. Table 4A-1 shows four alternative estimates of the U.S. demand schedule for energy and oil for 1980. If the price of energy were low between 1973 and 1980 ($3/bbl.), total energy consumption (in the central estimate) would be expected to increase by 28 percent, and total oil consumption by 50 percent. On the other hand, if the price were $11/bbl. during that period, demand for energy would be expected to rise 18 percent, while demand for oil would remain approximately constant. This clearly indicates that the price elasticity of oil demand is much greater than the price elasticity of energy demand.

The estimates made for U.S. energy demand (1980) in Table 4A-1 can be compared with other projections. The M.I.T. study[b] forecasts 1980 demand as shown in Table 4A-2.

The estimates are close to each other. The main difference is that our estimates in the $7/bbl. to $11/bbl. range are less sensitive to price.[c]

A similar comparison can be made regarding expected 1980 U.S. oil demand. The M.I.T. estimates in Table 4A-3 constitute the sum of the United States oil output and oil imports. The M.I.T. Judgmental projection is almost identical to our central estimate with $n = 0.3$. The M.I.T. model provides lower estimates.[d]

While the U.S. is self-sufficient in coal, it is not so in oil. In 1973 the U.S. imported 2.2 bil.bbl. of oil. The U.S. demand for oil *import* is the difference between its oil demand and supply schedules. The U.S. supply schedule for 1980 (estimated in Appendix 3A) rises from 2.5 bil.bbl. at $3/bbl. to 5.5 bil.bbl. at

---

[a]The parameters of the U.S. energy demand function are assumed identical with the world's function.

[b]Cf. Adelman et al., *Technology Review, op.cit.*

[c]Applying the parameter $n = 0.3$ (rather than $n = 1.0$) makes our estimate somewhat more sensitive to price. The central estimate, for example, varies between 15.8 bil.bbl. at $7/bbl. and 15.3 bil.bbl. at $11/bbl.

[d]Hudson and Jorgenson's respective projection at a price of $7/bbl. is 6.8 bil.bbl. compared to 6.9 bil.bbl. at our central estimate. (Cf. Hudson & Jorgenson, *op. cit.*, Table 8, p. 492).

Table 4A-1
Estimated U.S. Demand for Energy and Oil, 1980
(Bil. Bbl. Oil Equivalent)

Income Elasticity	0.75		1.00		1.00		1.25	
Price Elasticity	−0.10		−0.10		−0.20		−0.10	
Price ($/bbl.)	Energy	Oil	Energy	Oil	Energy	Oil	Energy	Oil
3	15.6	8.4	16.6	9.4	16.6	9.4	17.6	10.4
4	15.2	7.0	16.2	8.0	15.8	7.6	17.1	8.9
5	15.0	6.7	15.9	7.6	15.3	7.0	17.0	8.7
6	14.8	6.3	15.8	7.3	14.9	6.4	16.7	8.2
7	14.7	6.0	15.6	6.9	14.7	6.0	16.6	7.9
8	14.6	5.7	15.5	6.6	14.5	5.6	16.5	7.6
9	14.5	5.4	15.5	6.4	14.4	5.3	16.4	7.3
10	14.5	5.3	15.4	6.2	14.3	5.1	16.3	7.1
11	14.4	5.0	15.4	6.0	14.2	4.8	16.3	6.9
12	14.4	4.8	15.3	5.7	14.1	4.5	16.2	6.6

Table 4A-2
Comparative Energy Demand Forecasts, 1980
(Bil. Bbl.)

	M.I.T.		Author's Estimates	
Price ($/bbl.)	Hudson Jorgenson[a]	Judgmental	Central	Highest
7	16.1	16.6	15.6	16.6
9	15.5	16.6	15.5	16.4
11	14.8	16.6	15.4	16.3

[a]Based on Hudson, E.A. and Jorgenson, D.W., "U.S. Energy Policy and Economic Growth, 1975-2000," *Bell Journal of Economics and Management,* Autumn 1974, pp. 461-514.

Table 4A-3
Comparative Oil Demand Forecasts, 1980
(Bil. Bbl.)

			Author's Central Estimate	
Price ($/bbl.)	M.I.T. Model	Judgmental	n = 1.0	n = 0.3
7	5.9	7.1	6.9	7.2
9	4.5	6.1	6.4	6.4
11	3.4	5.4	6.0	5.6

$9/bbl., assuming constant price. Table 4A-4 shows the estimated U.S. demand for oil *import* in 1980.

The central estimate (n_p = -0.10, n_y = 1.00) shows that if the price remains low U.S. demand for oil import is expected to be *over three times greater* in 1980 than in 1973. Maintenance of a constant level of import requires a price of $8/bbl. Full independence is unlikely to be achieved at prices lower than $9/bbl. The alternative demand estimates differ somewhat from the central one but do not significantly change these conclusions.

Table 4A-4
Estimated U.S. Net Demand for Oil Import, 1980
(Bil. Bbl.)

Income Elasticity	*0.75*	*1.00*	*1.00*	*1.25*
Price Elasticity	*−0.10*	*−0.10*	*−0.20*	*−0.10*
Price ($/bbl.)				
3	5.9	6.9	6.9	7.9
4	4.0	5.0	4.6	5.9
5	2.2	3.1	2.5	4.2
5	1.6	2.6	1.7	3.5
7	1.2	2.1	1.2	3.1
8	0.7	1.6	0.6	2.6
9	−0.1	0.9	−0.2	1.8

5

The Price Policy Model

We now present our model for determining optimal price policy over time, based on the demand and supply behavior described earlier. This chapter shows the model in its basic form for OPEC as a whole. In the following chapters, the model will be applied both to OPEC and to separate country groups within OPEC.

The OPEC cartel is assumed to operate as a price-leader oligopoly. It determines its output as if it were a *monopolist* with regard to its net oil demand—the difference between total oil demand and non-OPEC oil supply, as shown schematically in Figure 5-1. The diagram is but a static description; it permits optimal price to be determined for only one period. In reality we face a dynamic problem.

Since present price affects future demand and competitive supply, the cartel cannot determine the optimum price unless it simultaneously solves the future price pattern over a fairly long period of time. This requires the use of a

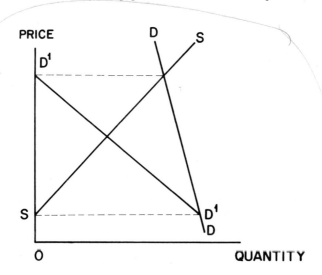

Note: $\overline{DD}$ = World oil demand; $\overline{SS}$ = non-OPEC oil supply; $\overline{D^1D^1}$ = net oil demand for OPEC oil.

Figure 5-1. Schematic Net Oil Demand

mathematical technique called *dynamic programming*. Our dynamic-programming model consists of the following components: world energy demand, world non-oil energy supply, and non-OPEC oil supply. The goal function is to maximize the net present value of future sales plus the reserves remaining at the end of the planning period. The oil reserves of the OPEC members constrain their price policy.

Demand

World demand for energy is given by the following function:

$$Q_t = \left\{ 1 + [(1 + g)^t - 1]\, n_y \right\} (Q_M + KP_t^{-n}) \qquad (5.1)$$

where t = number of years after 1973
       Q_t = amount demanded in year t (in terms of oil equivalents)
       P_t = price in dollars/bbl. in year t
       g = annual rate of income increase
       n_y = income elasticity of demand
       Q_M, K, and n = parameters of the 1973 demand function

Equation 5.1 is the demand function presented in Chapter 4. The righthand parentheses represent the basic demand in 1973 as a function of price. The lefthand braces represent the measurement of the income effect (i.e., the rate at which demand rises with respect to income). The income effect depends on the rate of income increase (g) and the income elasticity of demand (n_y).

Equation 5.1 can be rewritten with the specific 1973 quantity demanded (30.7 bil./bbl.) and price ($3/bbl.) by using Equation 4.2 as follows:[a]

---

[a] Equation 5.2 differs from 5.1 in its righthand braces. For 1973 this section in 5.1 is:

$$30.7 = Q_M + K3^{-n} \qquad (a.1)$$

and Equation 4.2 becomes:

$$n_p = -n\left(1 - \frac{Q_M}{30.7}\right) \qquad (a.2)$$

From (a.2) it follows that:

$$Q_M = 30.7\left(1 + \frac{n_p}{n}\right) \qquad (a.3)$$

$$Q_t = \left\{1 + [(1+g)^t - 1]\, n_y \right\}\left\{30.7 - 30.7\, n_p/n \; [(3/P_t)^n - 1]\right\} \qquad (5.2)$$

Q_t thus can be calculated for any combination of g, n_y, n_p, and n.

Supply

Non-Oil Energy Supply

This supply function is taken from Chapter 3:

$$S_t^1 = \begin{cases} (0.940 + 0.020P_{t-1})S_{t-1}^1 & \text{for } P \leqslant \$4 \\[2mm] (1.000 + 0.005P_{t-1})S_{t-1}^1 & \text{for } P \geqslant \$4 \end{cases} \qquad (5.3)$$

where S_t^1 = supply of non-oil energy in year t.

The function shows that the supply in year t depends on the supply in year $t-1$ and on a rate of change that depends on the price in year $t-1$.

Non-OPEC Oil

$$S_t^2 = \begin{cases} (0.899 + 0.027P_{t-1})S_{t-1}^2 & \text{for } P \leqslant \$5 \\[2mm] (1.010 + 0.005P_{t-1})S_{t-1}^2 & \text{for } P > \$5 \end{cases} \qquad (5.4)$$

where S_t^2 = supply of non-OPEC oil in year t.

The above function is also taken from Chapter 3. As in the preceding function, S_t^2 depends on S_{t-1}^2 and P_{t-1}.

Net Demand for OPEC Oil

The net demand for OPEC oil is the difference between total energy demand and the two other sources of energy supply. Let D equal net demand for OPEC oil. Thus:

$$D_t = Q_t - S_t^1 - S_t^2 \qquad (5.5)$$

Substituting for (a.1) and rearranging gives:

$$K = -3^n \cdot 30.7\, \frac{n_p}{n} \qquad (a.4)$$

Substituting (a.3) and (a.4) for (5.1) and rearranging gives Equation 5.2.

48

Since Q_t depends on P_t, and S_t^1 and S_t^2 depend on P_{t-1}, it follows that D_t depends on both P_t and P_{t-1}.

Goal Function

OPEC's goal function is defined as the maximization of the total wealth inherent in the OPEC oil reserves. Normally this definition would require that the planning period be long enough for all present oil reserves to be utilized.

However, it is impossible to make sound projections of long-term energy demand and competitive supply. Moreover, the effect of far-future demand and supply on near-term optimal prices is diminished by the time-discount factor. Therefore we limit our planning period to about 15 years. Oil prices for 1974 and 1975 are available; hence we apply our model to the 1976-1990 period.

Since the planning period is limited, the goal function must be divided into two components: the present value of net oil revenues during the planning period, and the present value of the oil reserves remaining after the planning period.

Present Value of Net Oil Revenues

The present value of net oil revenues from the sale of oil during the planning period is given by the following formula:

$$V_1 = \sum_{t=1}^{T} \frac{1}{(1+r)^t} (P_t - C_t)D_t \tag{5.6}$$

Where T is the horizon year of the planning period; $P_t - C_t$ is the net revenue/bbl. of oil in year t; and r is the discount rate for calculating the present value of future revenues.

Present Value of Remaining Oil Reserves

The present value of the reserves remaining after 1990 is calculated under the assumption that they will be sold off gradually over a 50-year period. Thus the reserves will be sold off by 2040, and the mean year is 2015. The present value of these reserves can be roughly estimated by discounting their value back from 2015 for about 40 years. This is given by the following formula:

$$V_2 = \frac{1}{(1+r)^{T+25}} (P^* - C^*) (R - \sum_{t=1}^{T} D_t) \tag{5.7}$$

where P^* and C^* are the representative price and cost per barrel of oil after the planning period; $(P^* - C^*)$ is the representative net revenue per barrel of oil after the planning period; R stands for total present reserves; and $\Sigma_{t=1}^{T} D_t$ is the total output of oil from these reserves during the planning period. $(T + 25)$ is the mean year during which the remaining reserves are sold. Therefore $1/(1 + r)^{T + 25}$ is the "average" present value of \$1 of the remaining reserves.

The present value of potential profits from present oil reserves is the sum total of V_1 and V_2. Equation 5.7, the value of the reserves remaining after 1990, can be divided into two components:

$$V_2 = \frac{1}{(1 + r)^{T + 25}} (P^* - C^*) R - \frac{1}{(1 + r)^{T + 25}} (P^* - C^*) \sum_{t = 1}^{T} D_t \quad (5.8)$$

The first component measures the present value of the remaining reserves if no portion of the presently available reserves (R) was sold during the planning period. This component is constant in the goal function, independent of the price pattern during the planning period; thus it can be deleted from the goal function without affecting the optimal solution.

The second component measures the loss in the value of the reserves remaining after 1990 due to the sale of oil during the planning period. In order to obtain revenue from the sale of oil during the planning period, the producer relinquishes the opportunity to sell this oil after the planning period. Therefore this component constitutes the opportunity cost of selling the oil during the planning period. It depends on the price policy during the planning period (which determines total sales, D_t), and therefore cannot be deleted from the goal function.

Thus we can simplify the goal function by deleting the first component of Equation 5.8. The balance will be referred to as the *net gain* from the sale of oil during the planning period. The gain is net because it measures the present value of net revenues from oil sold *minus* the present value of this oil if it had remained in the ground.

This presentation helps us understand how new discoveries of reserves by OPEC members may affect OPEC's price policy. Since the goal function is independent of the available reserves, new additional reserves do not affect it directly. However, greater reserves extend the duration of their utilization after the planning period; therefore the imposed assumption of fifty years utilization time for the remaining reserves should be adjusted. This indirect effect is expressed by increasing the $T + 25$ power in the $1/(1 + r)^{T + 25}$ discount factors. This somewhat reduces the opportunity cost of selling oil during the planning period, resulting in a new optimal solution with slightly greater sales at slightly lower prices.

Given a discount rate of 8 percent, the value of $1/(1 + r)^{T + 25}$ is 0.046. This means that the present value of each dollar received from post-1990 sales of the remaining reserves is only 4.6¢. Thus if we assume that the oil price after the planning period will be between \$15/bbl. and \$25/bbl. (in real purchasing power), the present value of the net revenue from one barrel of oil sold after the planning period is about \$1. Thus we assume that in Equation 5.8 the expression $1/(1 + r)^{T + 25} (P^* - C^*) = 1$. This further simplifies the presentation of the goal function in its net gain form.

The net gain function can now be presented as follows:

$$V_N = \sum_{t=1}^{T} \frac{1}{(1+r)^t} (P_t - C_t)D_t - \sum_{t=1}^{T} D_t$$

$$= \sum_{t=1}^{T} [\frac{1}{(1+r)^t}(P_t - C_t) - 1]D_t$$

(5.9)

where V_N is the net gain from oil sold during the planning period.

Whatever price policy is chosen affects the quantity of oil sold during the planning period. A change in price affects the net gain function in two ways: first, by changing net revenues per barrel during the planning period; second, by changing the quantity sold, D_t. A change in D_t has two effects: on total revenues during the planning period, and on the loss in the present value of oil after the planning period if it had remained in the ground. The second effect of a change in D_t is smaller than the first one. Therefore the price policy affects the goal function primarily through its effect on net revenues from sales during the planning period.

Appendix 5A

The Model

The dynamic-programming model used in this study is summarized below.

Let:
Q	=	amount of energy demanded
P	=	energy price (per barrel of oil equivalent)
C	=	cost of production (per barrel of oil)
g	=	annual growth rate of national income
r	=	interest rate
n_p	=	price elasticity of energy demand at $P = \$3/\text{bbl}$.
n_y	=	income elasticity of energy demand
R	=	OPEC's present proven reserves
D	=	OPEC sales
S^1	=	amount of non-oil energy supplied
S^2	=	amount of non-OPEC oil supplied
t	=	year
T	=	planning horizon
P^* and C^*	=	price and cost per barrel after the planning period
n	=	parameter of the energy demand function.

The formulation of the model is as follows:

$$V = \text{Max} \sum_{t=1}^{T} \frac{1}{(1+r)^t} (P_t - C_t)D_t + \frac{1}{(1+r)^{T+25}} (P^* - C^*)(R - \sum_{t=1}^{T} D_t)$$

subject to:

$$\sum_{t=1}^{T} D_t \leqslant R$$

$$W \leqslant P \leqslant V$$

where

$$D_t = Q_t - S_t^1 - S_t^2$$

$$Q_t = \left\{ 1 + [(1+g)^t - 1] n_y \right\} [Q_0 - Q_0 n_p / n(3^n P_t^{-n} - 1)]$$

$$S_t^1 = \begin{cases} (0.940 + 0.020P_{t-1})S_{t-1}^1 & \text{for } P \leqslant \$4 \\ (1 + 0.005P_{t-1})S_{t-1}^1 & \text{for } P \geqslant \$4 \end{cases}$$

51

$$S_t^2 = \begin{cases} (0.899 + 0.027P_{t-1})S_{t-1}^2 & \text{for } P \leqslant \$5 \\ \\ (1.010 + 0.005P_{t-1})S_{t-1}^2 & \text{for } P > \$5 \end{cases}$$

and Q_o = 30.7 bil.bbl. oil equivalent.

S_o^1 = 13.6 bil.bbl. oil equivalent.

S_o^2 = 5.7 bil.bbl. oil.

C_t = \$1.5/bbl.

The model solves the vector of $P_t(t = 1, \ldots, T)$ for any combination of g, n_y, n_p, and n.

The goal function can be simplifed by assuming that $1/(1 + r)^{T + 25}(P^* - C^*) = 1$ and by deleting R, which is a constant parameter. We thus have the net gain goal function as:

$$V_N = \text{Max} \sum_{t=1}^{T} \frac{1}{(1+r)^t}(P_t - C_t)D_t - \sum_{t=1}^{T} D_t = \sum_{t=1}^{T} [\frac{1}{(1+r)^t}(P_t - C_t) - 1]D_t$$

6

Price Policies for OPEC

In this chapter our price optimization model is applied to OPEC as a single co-herent body. We assume that OPEC behaves as a price-leading cartel with a well-defined goal function. However, this is not necessarily the only interpretation of the model. In Chapter 2 we described the usefulness of such a normative model to a positive approach in explaining past price development and possible future price patterns.

The analytical solution of the model provides a price vector for the period 1976 to 1990 that allows OPEC to achieve the maximum net gain from its oil reserves. We refer to this as the optimal price pattern. The specific solution ob-viously depends on the assumptions made regarding future demand and supply of energy. We examined a number of alternative demand and supply functions of the type appearing in Chapter 5 by changing the values of some of their parameters, and found that the pattern of the optimal solutions repeats itself. For the sake of simplicity we therefore show in this chapter only one set of findings. This set is based on supply functions with the specific parameters as shown in Chapter 5 and an energy demand function where the income elasticity demand (n_y) is 1.00, the price elasticity of demand at a price of \$3/bbl. (n_p) is -0.1, the parameter n is 0.3, and the annual rate of income growth (g) is 0.04. All the projections are calculated in terms of constant 1974 purchasing power and refer to the 1976-1990 time period.[a] The following results should indeed be regarded merely as indicative examples.

First, we test some intuitively selected price patterns that are often assumed in current literature. This is done by incorporating the price patterns into the model and then checking the resulting flow of oil revenues and the value of the goal function. For the sake of simplicity, we have limited our choice patterns to (1) constant prices over time and (2) prices increasing at a constant rate-per-time period.

Constant Prices

Let us suppose that OPEC is able to determine price policy only once—so that

---

[a]The original form of the model presented in Chapter 5 is based on 1973 data. In order to up-date it to 1975 the following constraints were added: $P_{1974} = P_{1975} = \$11/\text{bbl}.$; $S^1_{1975} = 13.6$ bil. bbl. oil equivalent; $S^2_{1975} = 5.7$ bil. bbl. oil.

this price will remain constant (in real terms) at least until 1990. Thus the price will be automatically adjusted to the rate of world inflation. We have examined the implications for OPEC of price levels from $4/bbl. to $20/bbl.

Estimated Demand

The effect of selected price levels on demand for OPEC oil is summarized in Table 6-1. If the price remains low (e.g., $4/bbl.), the demand for OPEC oil will increase very rapidly (at a rate similar to past performance). The higher the price becomes, the more slowly the demand increases. At $10/bbl. demand stabilizes at its 1973 level. At prices higher than $10/bbl. demand decreases. This decrease accelerates with time—so that at prices higher than $12/bbl. demand virtually disappears by 1990. This finding indicates that constant real prices of $13/bbl. and over could not persist through the 1980s.

Table 6-1
Estimated Demand for OPEC Oil at Constant Prices: 1980, 1985, 1990
(Bil. Bbl.)

Price ($/bbl.)	1973[a]	1980	1985	1990	Annual Rates of Increase (%) 1973-1980	1980-1990
4	11.4	18.5	25.4	34.0	7.2	6.3
6	11.4	15.2	19.4	24.7	4.2	5.0
8	11.4	13.1	15.5	18.3	2.0	3.4
10	11.4	11.3	11.8	11.7	−0.1	0.4
11	11.4	10.5	9.9	8.3	−1.2	−2.2
12	11.4	9.7	8.1	4.8	−2.3	−6.7
13	11.4	8.9	6.2	1.1	−3.5	−18.6
14	11.4	8.1	4.3		−4.7	

[a]The figure for 1973 represents actual demand and serves as a basis for comparison.

Estimated Net Revenues

OPEC's estimated net revenues for each constant price level are derived by multiplying OPEC's demand by the associated net price. Net revenue estimates for 1980, 1985, and 1990 appear in Table 6-2. The table shows that for 1980, the higher the price, the greater the real value of net revenues. However, for 1985 and 1990 the pattern changes: revenues increase with price up to the $8/bbl. to $9/bbl. range but decrease at higher prices.

Table 6-2
Estimated OPEC Net Revenues at Constant Prices: 1980, 1985, 1990
(Billions of Dollars)

Price ($/bbl.)	1980	1985	1990
4	46	64	85
6	68	87	111
8	85	100	119
9	91	102	112
10	96	100	100
11	99	94	79
12	101	85	51
13	102	71	13
14	102	54	

Estimated Net Gain

Estimated OPEC net gain is derived by calculating the value of the goal function at each of the constant price levels. See Table 6-3. The net gain (column 4) is the difference between the present value of the net revenues from sales (column 2) and the opportunity cost of the oil sold if it had been left in the ground (column 3). Since we evaluate the oil reserves after 1990 at a present value of $1/bbl. (see Chapter 5), the numbers in column 3 also represent the amount of oil sold (in bil. bbl.) during the planning period.

The highest net gain for OPEC is achieved at a price of $11/bbl. Indeed the highest present value of revenues from sales is achieved at a price of $10/bbl.; however, due to the greater amount of oil sold, the net gain at $10/bbl. is somewhat lower than its maximum. Net gain becomes insensitive to small price changes near the optimal price level; any price from $10/bbl. to $12/bbl. yields about the same net gain.

It is quite surprising to find that in its first test the model yields an optimum price that is quite near the actual level determined by OPEC as of January 1974. This result depends on the combination of assumptions incorporated into the model—e.g., the price elasticity of demand, the reaction of competitive supply, the income elasticity of demand, the expected rate-of-income increase, and the discount rate. Sensitivity analysis, however, indicates that the optimum price is only slightly affected by changes in these parameters—except for one: the reaction of supply to price level. Alternative assumptions regarding this factor significantly affect the constant price at which maximum net gain is reached. It must be remembered that the supply-reaction function is subject to great uncertainty. Since past prices never reached the present high level, no past data can aid in reducing the uncertainty.

Table 6-3
Estimated OPEC Net Gain Under Constant Prices[a]
(Billions of Dollars)

Price ($/bbl.) (1)	Present Value of Net Revenue (2)	Opportunity Cost of Oil Sold (3)	Net Gain (4)
4	550	357	193
5	649	308	341
6	744	280	464
7	819	254	565
8	874	230	644
9	909	206	703
10	925	183	742
11	919	159	760[b]
12	893	136	757
13	845	112	733
14	775	88	687

[a]Present values are calculated at 8 percent interest rate.
[b]Indicates the maximum value.

The estimates for this study were prepared on the basis of available information. They were completed before any analysis was made of the data. Thus the resulting "optimal" constant price level should be interpreted with great care. Interestingly, however, the projected sales of OPEC oil are quite consistent with several other projections.[b]

Prices Increasing at a Constant Rate

An analysis of optimal pricing over time (as summarized in Chapter 2) leads to consideration of a price pattern which rises with time. In both purely competitive and monopolistic market structures, as we have seen, the optimal price pattern is one under which the net price rises at a constant rate. This rate is determined by the market rate of interest. We have accordingly examined a set of alternative price patterns in which the net price rises at various rates. Some of the patterns which provide the highest net gain are summarized in Table 6-4.

The greatest net gain is achieved at an initial price of $7/bbl. with an 8 percent net price increase. This policy results in the highest present value of net

[b]See Appendix.

Table 6-4
Estimated OPEC Net Gain under Gradually Rising Prices
(Billions of Dollars)

Initial Price ($/bbl.)	Annual Rate of Net Price Increase	Present Value of Net Revenue	Opportunity Cost of Oil Sold	Net Gain
7	6	954	201	753
7	7	963	191	772
7	8	963	180	783[a]
8	4	947	191	756
8	5	954	181	773
8	6	952	170	782

[a]Maximum value.

revenues and the second lowest opportunity cost of oil sold. Other alternatives (lower and higher initial prices and different rates of net price increase) are all found to be inferior to the policy of $7/bbl. at 8 percent.

Under this policy, the price rises gradually to $17.65/bbl. after 15 years. OPEC's net gain ($783 bil.) appears to be greater than that provided by the "optimal" constant price of $11/bbl. ($760. bil.).

Optimal Price Policy

The optimal price policy is determined by a solution of the dynamic-programming model. Computer constraints made it necessary to solve the problem with a limiting time horizon of 15 years; this limit introduced a bias into the solution. Since the pricing of the post-planning period is not solved by the model but is assumed exogeneously, the optimal price in the later years of the planning period may increase sharply in order to reap maximum gains during the planning period—disregarding its post-horizon effects on competitive supply. Consequently, price bounds must be introduced to offset the bias.

Two sets of price bounds were chosen that agree with the best patterns produced by the above analysis. The first set has a lower limit of $4/bbl. and an upper limit of $17.65/bbl.; the lower bound was selected in order to allow for an alternative of a low-price policy. The upper bound represents the maximum price reached under the best gradually rising price policy. The second set has bounds of $3/bbl. and $11/bbl. to reflect the range of the 1973–1974 price hike.

The optimal solution under each price bound alternative has a consistently repeating pattern. In the early period the price coincides constantly with the lower bound; in the later period the price jumps to the upper bound and remains there until the horizon. This corner solution pattern repeats itself with various bounds and horizons and with various parameters of the supply and demand functions.[c] The solutions differ in the values of lower and upper prices and in the year of the price jump. We refer to this price pattern as a *one-shot price hike.*

The length of the planning horizon greatly affects the optimal price pattern. A sensitivity analysis shows that the shorter the horizon the earlier the price hike should be. With bounds of $4/bbl. and $17.65/bbl., for example, the optimal pattern is as follows: for a 15-year horizon (1976–1990) the price jumps in 1982. For a 13-year horizon the price is hiked one year earlier. For an 11-year horizon the price jumps three years earlier. For a very short horizon the price jumps at the starting point (1976).

A sensitivity analysis was made for a 15-year planning period as to the year in which the price is hiked. The optimum year for the price hike is 1982 with a net gain of $822 billion. Hiking the price earlier or later results in a smaller net gain: 1979—$712 billion; 1985—$731 billion. However, near the optimum year, net gain has a low sensitivity to the year of the price hike. A hike in 1981 and 1983 yields the same net gain—$810 billion.

A similar analysis for the $3/bbl. and $11/bbl. price bounds shows that the optimum year of price hike is 1978 with net gain of $799 billion. Hiking the price to $11/bbl. at the beginning of the period (which coincides with the optimum *constant* price) yields a lower net gain of $760 billion.

Table 6-5 summarizes OPEC's net gain under the main price policies. The conclusions that can be drawn from the table are:

1. The best constant price policy ($11/bbl.) is inferior to any other policy.
2. The best one-shot price hike ($4/bbl. to $17.65/bbl. in 1982) is better policy than the best gradually rising price (from $7/bbl. at 8 percent). The net gain of the former is $822 bil., that of the latter is $783 billion.

Within the one-shot price-hike policy, the more the price bounds are relaxed, the greater is the net gain. This was substantiated by a sensitivity analysis.

Simulation of Past Oil Price Development

The one-shot price-hike pattern recurred consistently as the optimal price policy in a great number of sensitivity analyses. This pattern is, in fact, a description of the history of oil prices. After a long period at $2/bbl. to $3/bbl.

---

[c]Also appearing (albeit infrequently) were interim prices over a period of 1 to 2 years, indicating a step-wise price hike from the lower to the upper bound. These interim prices exclude the possibility that the model is based on convex functions.

Table 6-5
Comparison of Price Policies for OPEC
(Billions of Dollars)

Price Policy	Present Value of Net Revenues	Opportunity Cost of Oil Sold	Net Gain
Constant Price of $11/bbl.[a]	919	159	760
One-Shot Price Hike from $3/bbl. to $11/bbl. in 1978	1020	221	799
Gradually Rising Price from $7/bbl. at 8%	963	180	783
One-Shot Price Hike from $4/bbl. to $17.65/bbl. in 1982	1033	211	822

[a]Also can be perceived as a one-shot price hike with the hike occurring at the outset.

the price was hiked to $11/bbl. to $12/bbl. within less than three months between October 1973 and January 1974. Thus we find that the normative optimization model becomes a simulation model, i.e., a model that simulates the price pattern that actually occurred.

Given the basic parameters in the dynamic functions, the extent of the price hike and its optimal timing depend on the price bounds and on the length of the planning period. We reconstructed the model with historical data to see if the 1973-1974 price hike could have been foreseen on the basis of data available before 1973.

We chose 1970 as the reference year. Total energy demand that year was the equivalent of 26.6 bil.bbl. of oil. The non-oil energy supply was the equivalent of 12.9 bil.bbl. The non-OPEC oil supply was 5.6 bil.bbl. and OPEC's output was 8.1 bil.bbl. Taking these figures and applying the same dynamic model under the same alternative price bounds, the resulting optimal price pattern is, again, a one-shot price hike.

We checked the optimal year for a price hike—at the different price bounds—for OPEC as a whole and for individual subgroups. The optimal year for a hike from $3/bbl. to $11/bbl. for OPEC as a whole is 1975. For a hike from $4/bbl. to $17.65/bbl. the optimal year is 1978—yet the net gain is somewhat smaller.

In reality OPEC hiked the price to about $11/bbl. one to two years earlier than the "optimal" year indicated by the model. At first glance this slight difference might seem inconsistent with the model.

The above solution, however, is based on the assumption that OPEC is a coherent cartel in which all partners share a common interest regarding optimal price policy. This assumption is subject to critical analysis. In the next chapter it will be demonstrated that conflicts of interest over optimal price policy do exist among individual members and groups of members in OPEC. It will be seen that the optimal year for the price hike differs among the groups. We will then be able to interpret OPEC's unified price policy as the outcome of conflicting interests within OPEC. The use of the model for the simulation of actual price developments will then be more complete.

7

Conflicting Price Policies within OPEC

We now examine OPEC not as a single coherent body but as an organization of members with different shares in the oil market and different oil reserves. We divide the OPEC members into four groups: Saudi Arabia, Iran, other Middle East members, and non-Middle East members. This permits us to uncover conflicts among the groups with regard to price policy.

The OPEC goal function of net gain maximization is applied to each group separately. The net gain for each group is calculated under different price patterns. This procedure reveals the optimal price patterns for each group and shows whether they are consistent or conflicting. These findings serve as a basis for policy analysis and for political implications.

Determination of Market Share

The higher the oil price, the smaller is the long-run demand for OPEC oil. Since the output capacity of the OPEC members exceeds their sales, a method of output rationing and market share allocation must be devised and accepted by the members. If an acceptable market sharing method is not found, OPEC may break down; the excess output of a nondisciplined OPEC members will push oil prices downward. On the other hand, an acceptable market sharing method may still cause conflicts about price; countries with a low ratio of output to reserves may prefer a lower price than do countries with a higher ratio. This problem, which affects the very existence of OPEC, is analyzed below.

The model determines the market share of each OPEC member for any year by the following procedure.

Let R_t = total OPEC reserves in year t
$\quad Y_t$ = total OPEC output in year t
$\quad j$ = country

Thus R_t^j = reserves of country j in year t
$\quad Y_t^j$ = output of country j in year t

Given these parameters for any year t, each country's share in year $t + 1$ is *assumed* to be determined as follows:

61

$$Y^{j}_{t+1} = \begin{cases} Y^{j}_{t} + (Y_{t+1} - Y_{t})\dfrac{R^{j}_{t}}{R_{t}} & \text{for } Y_{t+1} > Y_{t} \\[2em] Y^{j}_{t} + (Y_{t+1} - Y_{t})\dfrac{Y^{j}_{t}}{Y_{t}} & \text{for } Y_{t+1} < Y_{t} \end{cases} \tag{7.1}$$

This rule is interpreted in the following manner.

If total OPEC output rises, each member receives an incremental share in the total incremental output proportionate to its share in the total oil *reserves*.

If total OPEC output declines, each member reduces its output by an equal proportion—the proportion by which total OPEC output declines.

This means that in periods of increasing total sales members with lower-than-average output/reserves ratios increase their market share; members with higher-than-average output/reserves ratios reduce their market shares. In periods of decreasing sales each member reduces its sales by the same proportion—so that the market share for each country remains intact.

The actual 1974 ratio of output to reserves was fairly uniform for most members—except for Saudi Arabia. Saudi's reserves strikingly exceeded those of any other country, but its output lagged behind its share in total reserves. Consequently, Saudi's ratio of output to reserves was much lower than that of other OPEC members.

Forecasts made before October 1973 predicted a rapid increase in oil demand and output and allocated a very substantial share in the incremental market to Saudi Arabia. It was expected that Saudi would at least double its output between 1973 and 1980. Since Saudi Arabia had about 24 percent of the total OPEC output in 1973, it was expected to have over 50 percent of the incremental market—and reach a 1980 market share of about 40 percent of OPEC output.

These projections seem to hold only for low prices. In Chapter 6 it was shown that the higher the price, the smaller is the rate of increase in demand for OPEC oil. At a price of about $10/bbl. demand for OPEC oil is expected to remain constant. At still higher prices demand will decline. Thus at $11/bbl. to $13/bbl. Saudi Arabia will not be able to increase its market share.

Using the market sharing procedure of Equation 7.1, we have calculated the Saudi Arabian share in the total OPEC market under alternative constant prices in Table 7-1.

Saudi Arabia's share of the OPEC market, which was less than 24 percent in 1973, is expected to have risen to over 40 percent in 1980 and 50 percent in 1990 if the price remained low. The higher the price becomes, however,

Table 7-1
Saudi Arabia's Estimated Share in the Total OPEC Market at Constant Prices, 1973-1990
(Percentage)

Price ($/bbl.)	1973	1980	1990
4	23.7	41.1	51.0
6	23.7	36.3	46.5
8	23.7	32.0	40.8
10	23.7	27.2	28.8
11	23.7	24.4	24.4
12	23 7	23.9	23.9

the smaller is the increase in Saudi's share. At $12/bbl. and above, its share does not increase but rather remains constant.

If Saudi's market share increases, the shares of other OPEC members decline; excluding Saudi Arabia, however, their shares relative to each other remain almost constant.

Given the market share of each member (or group of members), we can now calculate and compare its net gain under alternative OPEC price policy.

Optimal Price Policy for Individual Groups within OPEC

As long as the effect of price on demand differs from one member to another, the price level that maximizes net gains may also vary accordingly. This is indicated in Table 7-2, which summarizes the net gains of the four separate groups within OPEC under a constant price policy.

The table shows that $11/bbl. to $12/bbl. is the constant price range under which the net gain is maximized for three of the groups (Iran, other Middle East, and other OPEC). This is also the price range that maximizes the net gain for OPEC as a whole. Standing in contrast to this position is Saudi Arabia, which reaches its maximum net gain at $9/bbl.

Because of the size of its oil deposits, Saudi Arabia will be selling off its reserves remaining after the planning period farther into the future than will the rest of OPEC. If we assume an average time lag of about 10 years for the disposal of Saudi Arabian reserves, the value of 1 bbl. of its reserves remaining after the planning period is less than 50 percent of the value of 1 bbl. of the other members' remaining reserves. Imputing such a coefficient to Saudi's opportunity cost of oil sold provides the net gains that appear in Table 7-3.

Both Tables 7-2 and 7-3 reveal an inherent conflict of interest between Saudi Arabia and the other members of OPEC. While the optimal constant

Table 7-2
Estimated Net Gains under Constant Prices by Country Groups
(Billions of Dollars)

Price ($/bbl.)	Total	Saudi Arabia	Iran	Other Middle East	Other OPEC
4	193	62	33	56	42
5	340	120	55	96	69
6	464	163	75	131	95
7	565	192	93	162	118
8	644	207	110	187	140
9	704	210[a]	125	209	160
10	742	200	138	226	178
11	760[a]	185	146	237	192
12	757	181	147[a]	237[a]	192[a]
13	733	175	142	230	186
14	687	164	133	215	175

[a]Maximum value.

Table 7-3
Estimated Net Gain for Saudi Arabia under Constant Prices with Lower Opportunity Cost

Price ($/bbl.)	Net Gain ($ bil.)
4	141
5	184
6	218
7	239
8	247[a]
9	242
10	225
11	205
12	197
13	189
14	175

[a]Maximum value.

price policy for all other members is $12/bbl., that for Saudi Arabia is only $8/bbl. to $9/bbl. Saudi Arabia's maximum net gain (Table 7-3) is $247 billion at a price of $8/bbl. It is only $205 billion at OPEC's optimal price of $11/bbl.—the same net gain that Saudi can achieve at half the price ($5/bbl. to $6/bbl.). Thus Saudi Arabia is found to be indifferent to prices as high as $11/bbl. to $12/bbl. and as low as $5/bbl. to $6/bbl.

The same type of conflict is found under other price patterns. With gradually rising prices, the best policy for OPEC as a whole and for OPEC members excluding Saudi Arabia is $7/bbl. with an 8 percent annual increase. For Saudi Arabia alone, the best policy is $6/bbl. with a 8 percent annual increase.

With the one-shot hike pattern, Saudi's optimal year for hiking the price at any price bounds is consistently later than that of non-Saudi OPEC.

Thus it can be concluded that Saudi Arabia deviates from OPEC on optimal price policy. This will be examined further in the following section, which analyzes the 1973–1974 price hike.

Simulation of Price Policy for OPEC Subgroups

We have shown that the dynamic-programming optimization model can also serve as a basis for a simulation of past price behavior (see Chapter 6). We applied 1970 oil market data to the model and examined the resulting price pattern for the post-1970 years. However, the analysis was limited to OPEC as a coherent body.

We now repeat the procedure—but this time with OPEC divided into subgroups. The results indicate the same general characteristic found in the previous section of this chapter: while three of the four subgroups showed a consistent optimal price structure, Saudi Arabia's optimal pricing is different. For this reason, the following discussion presents Saudi Arabia versus the other members of OPEC.

The optimal solution is, again, the one-shot price-hike pattern. This applies to both Saudi Arabia and the rest of OPEC. However, the two groups differ as to the preferred price bounds and the optimal price-hike years for given price bounds. This is revealed in Table 7-4. The table provides interesting insight into the price policy alternatives confronting OPEC in 1970 and the actual choice of price bounds and price-hike year.

Price Bounds

For OPEC as a whole the $4/bbl. to $17.65/bbl. and the $3/bbl. to $11/bbl. bounds provided about the same maximum net gain ($444 billion versus $449 billion). However, an examination of the OPEC subgroups indicates substantial

Table 7-4
Estimated Net Gain by Country Groups: A View from 1970
(Billions of Dollars)

Price Bounds ($/bbl.)	Year of Price Hike	Saudia Arabia	All Other OPEC	Total OPEC
3-11	1973	117	318[a]	435
	1975	158	291	449[a]
	1977	169[a]	246	415
4-17.65	1978	172	272[a]	444[a]
	1979	177[a]	264	441

[a]Maximum value.

differences. On one hand, the highest net gain for Saudi Arabia is reached at the $4/bbl. to $17.65/bbl. policy ($177 billion versus $169 billion at its best $3/bbl. to $11/bbl. policy). On the other hand, all other OPEC countries benefit more from the $3/bbl. to $11/bbl. policy because it gives them a maximum combined net gain $46 billion greater than the $4/bbl. to $17.65/bbl. policy ($318 billion versus $272 billion).

Therefore the actual price hike to about $11/bbl. indicates that OPEC price policy was determined by the interests of the other OPEC members rather than by Saudi Arabia.

Year of Price Hike

Once the $3/bbl. to $11/bbl. bounds are selected, there is still conflict about the optimal year for the price hike. For Saudi Arabia it is 1977; for the rest of OPEC it is 1973. The best price-hike year for OPEC as a whole—1975—simply results from the combined net gain of all OPEC members.

Theoretically, if the power of determining price policy had been with Saudi Arabia, prices would not have risen in 1973-1974—but a few years later. Moreover, if price determination reflects compromise between conflicting interests, the price hike should have taken place around 1975. The fact that the prices were hiked earlier indicates that OPEC price policy was determined without regard to Saudi's interests and the Saudi nevertheless complied with the wishes of the other OPEC members.

The foregoing analysis illuminates the role of covert economic forces within OPEC that might explain real events. It also reinforces the argument that there is an inherent economic conflict between Saudia and the rest of OPEC. In addition, it indicates that price policy is affected not only by economic interest but also by political power. We now briefly review such political factors and their interconnection with economic considerations.

Political Implications within OPEC

The preceding economic analysis is an appropriate preface to a discussion of political implications. It indicates that the apparent cohesion of the OPEC members might be only a facade. The basic economic interests of Saudi Arabia actually deviate from those of the rest of OPEC. Thus Saudi Arabia faces a basic dilemma: cooperation with OPEC has economic disadvantages, but breaking the cartel may involve great political risks.

Cooperation with OPEC entails the following disadvantages for Saudi Arabia:

1. If Saudi Arabia sells its oil at OPEC's optimal price level, its net gain from this sale will be less than if it sold oil at a lower price. At Saudi Arabia's optimal constant price ($8/bbl.) its net gain would be $247 billion. At the OPEC optimal price ($11/bbl.) Saudi's net gain would be only $205 billion. Saudi can also achieve a net gain of $205 billion at the much lower rate of $5/bbl. to $6/bbl. Economically, therefore, Saudi Arabia is quite indifferent to whether the price policy is a high $11/bbl. or a low $5/bbl. to $6/bbl. and obviously prefers $8/bbl.
2. The OPEC price policy ($11/bbl.) is more advantageous to Saudi Arabia's potential rivals, particularly Iran. This policy gives Iran greater oil revenues for domestic investment, thereby increasing its rate of economic growth. It also provides Iran with more funds for military expenditure. The threat of an economically and militarily powerful Iran is too great for Saudi Arabia to ignore.
3. In response to Iran's increasing power, Saudi Arabia is rapidly expanding its army. The Saudi regime is forced to fill newly created, influential military positions with people whose loyalty may be questionable, thereby threatening the regime's stability.
4. Because of mutual economic and political interests with Western countries, the Saudi regime is concerned with their economic well-being. Saudi's adherence to OPEC's high price policy weakens the Western economies, thereby undermining Saudi's own goals.

Despite these disadvantages of cooperating with OPEC, Saudi Arabia may find it difficult to break away.[1] The present Saudi regime gained support and political legitimacy from other Arab countries by cooperating with OPEC on the use of oil as a political weapon. Any retreat from this cooperative stance will endanger the central leadership and prestige among Arab nations that Saudi Arabia has achieved. A break with OPEC is bound to incur great external political pressure on Saudi and give rise to internal instability.

This predicament prevents Saudi Arabia from taking an active role in reducing oil prices despite the clear advantage in doing so. It explains the ambiguity in Saudi's position in the 1974 and 1975 OPEC price-policy negotiations and sheds light on Saudi's repeated attempts to hold down the rate of OPEC's price increase.

Implications for Western Policy

An understanding of the economic and political factors affecting Saudi Arabia may help the Western world (particularly the U.S.) pursue a more effective political program for price reduction. The program should increase Saudi Arabia's economic motivation to lower the price and reduce the political risks to Saudi inherent in a policy of price decline.

There are several economic incentives that may encourage Saudi Arabia to reduce prices.

Protection against Inflation

Most of Saudi's revenues are surplus, accumulating in foreign investments.[2] For this reason Saudi Arabia is bound to suffer from world inflation more than any other OPEC members. As long as the accumulation of financial assets is not exceedingly large, the absolute value of this loss is small relative to the oil revenues. As the surpluses increase with time however, the loss from inflation will increase proportionately—becoming a crucial factor. When Saudia's accumulation of assets amounts to $200 billion for example, inflation of 10 percent will result in a real loss of $20 billion annually—approximately the same amount as net revenues. The West could offer indexed securities to protect Sandi's assets against inflation in return for a significant cut in oil prices. While this would involve a financial cost to the West, the cost would be a fraction of savings from a reduction in oil prices. For example, linkage guarantees at a 10 percent yearly inflation rate would require an initial annual commitment of $2 billion to $3 billion. This amount would gradually increase in the following years, reaching $20 billion in the 1980s (when Saudi's accumulated capital in this form would be about $200 billion). On the other hand, a price reduction from $11/bbl. to $6/bbl. or $7/bbl. would yield an annual saving of about $40 billion for the West. Thus the net gain for the West would be substantial.

Better Protection against Business Risk

The West could create new types of financial investment oppotunities designed to reduce investment risks inherent in existing securities.

Increased Benefits from Domestic Investments

If the West increases technical assistance for the development of Saudi Arabia's economy, internal investment opportunities for its oil revenues would increase. This assistance would provide know-how, patents, training, market connections, and managerial skills now seriously lacking in Saudi Arabia. The program could be developed with the help of multinational corporations and the support of Western governments. It would have to be carried out more rapidly and more effectively than were similar programs in the past.

Minimum Price Guarantee

Saudi Arabia may have apprehensions that if it cooperates in a policy of price decline, an uncontrolled reaction by other OPEC members might drop the price of oil *too* low. To alleviate this fear, the West could guarantee Saudi Arabia a minimum level of oil price by contracting conditional future sales with long horizon periods.

These and other measures may provide strong economic encouragement to Saudi Arabia to cooperate with the West in an oil price reduction. In addition, however, measures to reduce Saudi's political risks also seem to be necessary for such cooperation.

Reliable Guarantees of Political Survival

The most important political consideration is the guarantee of survival of the present Saudi regime. This is just as important to the West as it is to the regime since Saudi's stability is a prerequisite to the steady flow of oil.

The main problem lies in the nature and reliability of such guarantees. Political and military factors are involved, including the West's capability for military intervention in case of a coup d'état. Furthermore, because of Saudi Arabia's relatively new and precarious position as a major leader in the Arab world, any steps taken to reduce oil prices should be artfully executed so as to disguise Saudi's role—in order to avoid antagonizing the Arab countries and the OPEC. For example, an oil surplus within OPEC is expected because of the lagging demand (resulting from the oil price increase) and the increased production capability of most OPEC members. This provides the West and Saudi Arabia with an opportunity to reduce prices while tactically obscuring any cooperation, thereby avoiding implicating Saudi Arabia. However, this is strictly a political reflection and could be elaborated elsewhere.

Notes

1. The following discussion is based on: M. Abir, "The Role of Persian Gulf Oil in Middle East and International Conflicts," unpublished memorandum, 1975.

2. See estimates in Chapter 10.

Part 2

Oil Revenues and Economic Development

8

Utilization of Oil Revenues: A Conceptual Framwork

Introduction

In the first part of this book we explored possible future developments in OPEC oil demand and prices. In the process, we estimated the net oil revenues of OPEC members. These revenues are expected to increase both their domestic economic growth and their investments in foreign countries (hereafter referred to as "foreign investments" or "foreign capital"). We treat oil revenues as a source of national income—singled out in the list of sources but integrated into total national income when allocated to the main economic uses.

As we trace possible uses of oil revenues, our discussion will be limited to the main OPEC members—Saudi Arabia, Kuwait, other Persian Gulf Emirates, Iran, Iraq, and Libya. This group will be referred to as the Middle East OPEC members. We will concentrate on these countries for a number of reasons. First, they hold almost 90 percent of OPEC's total reserves and produced 75 percent of OPEC's total output in 1973; this share was expected to exceed 80 percent if the price remained low and total output increased rapidly. Second, oil revenues constitute a major sector of these countries' economies; therefore OPEC's price policy affects them greatly. Third, these countries are closely related by economic and political ties because of their geographical proximity. The other OPEC members (Indonesia, Algeria, Nigeria, Venezuela, and Equador) were excluded by the above criteria.

This Part is divided into three chapters. Chapter 8 presents the conceptual framework. Chapter 9 describes a model of economic growth and capital accumulation for the countries. Chapter 10 summarizes the model's main projections of economic growth and foreign capital accumulation.

Criteria for Revenue Allocation

The oil revenues can be allocated to four main economic uses:

1. Increased private and public consumption in order to raise the standard of living and public welfare.
2. Increased throwaway spending (i.e., spending that increased neither public welfare nor production capacity)—for example, excessive military spending and "monumental" projects. This spending is made possible by the vast amount of easy-come income and constitutes waste.

73

3. Increased domestic investment for economic development and growth.
4. Increased investment in foreign countries.

In order to rationalize the allocation of oil revenues among these four uses, we assume the following set of goals:

1. To develop the economy through investments that provide the highest expected return for any given level of economic and political risk
2. To increase the domestic standard of living
3. To increase national security
4. To maintain and strengthen the stability of the regime
5. To increase the country's international political power

We first explore the allocation of oil revenues between consumption and investment. After a brief discussion of throwaway spending, we examine the allocation of investment between domestic and foreign uses.

Consumption versus Investment

Increasing the level of consumption raises the standard of living. One could argue, therefore, that the oil revenues should be used primarily to increase current consumption to a maximum level. However, increased consumption does not efficiently increase the level of achievement of the other goals.

The oil revenues are expected to flow only for a limited and uncertain time period. If part of the revenue is not saved and invested, consumption will fall drastically as soon as the flow of revenue drops. Moreover, the increased oil revenues cannot be consumed immediately, so there are limits on the rate of increase in consumption. In many of these countries the population is small relative to the increase in oil revenues; an extraordinary per-capita level of consumption would be required to absorb the revenues. Increased consumption requires far-reaching changes in private and social behavioral patterns, which take time to develop.

Consumption changes as a function of the change in normal income (permanent income) rather than as a direct result of windfalls. In order to increase consumption at a greater rate, a government could increase transfer payments and gifts to the population—but such a policy could undermine the country's social and economic viability. For all these reasons it is expected that only a fraction of the increased oil revenues will be directly allocated to increased consumption.

A better way to increase the standard of living is by investments that generate permanent future income rather than by immediate spending on imports of goods and services. The return on these investments would be utilized partly for a gradual increase in consumption and partly for further investment and accumulation of capital.

In light of the tremendous increase in oil revenues and the current limits on consumption, it is expected that most of the increased revenues will be invested and only a small proportion will be consumed immediately. A schematic representation of this process appears in Figure 8-1. The chart shows the utilization over time of a one-time influx oil revenue. The effect of subsequent revenue inflows will be cumulative; each new influx will add (in the same schematic structure) to the flow of revenue and investment returns.

Throwaway Spending

Throwaway spending contributes nothing to economic growth or an increase in the standard of living. It might be "justified" by a presumed contribution to national security or the strengthening of the regime. But compared to productive investments, this spending is largely inferior and therefore should be eliminated. The fact that such spending does exist makes it necessary for us to take it into consideration through an arbitrary procedure. In the model presented in the following chapter we allocate a predetermined proportion of the oil revenues to this use.

Domestic Investments versus Foreign Investments

The question is frequently raised as to whether domestic investment or foreign investment is more efficient in achieving OPEC's goals. Posing the question in this matter actually introduces a bias into the analysis—because *both* investment strategies contribute to goal achievement. The proper question to raise is what *proportion* of total investments should be allocated to each type in order to maximize goal achievement. We discuss this problem with regard to each of the five main goals.

Investment Return and Economic Risk

Expected return and risk are generally considered to be major criteria of business investment. If there were no risk involved, the business goal would be simply to select the investment opportunities that provide the highest return. In reality, however, investments are made under uncertain conditions. It is generally accepted that investors are risk-averse; they choose the investment opportunities entailing the lowest risk at a given level of *expected* return.

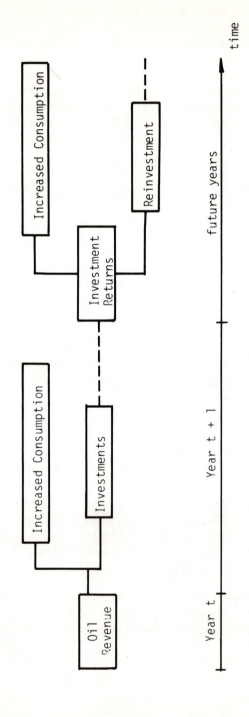

Figure 8-1. Flowchart of Oil Revenue Utilization

Investors facing uncertainty can reduce their investment risk by diversification, i.e., by investing in a number of opportunities thereby creating a mixed portfolio. Any given level of expected return is associated with a minimum risk that can be achieved by a proper diversification. A portfolio that achieves such a minimum risk level for a given level of expected return is called an *efficient portfolio*. The higher the expected return, the greater is the minimum level of risk attained by the respective efficient portfolio. Thus returns must be weighed against risks in order to make the proper choice among efficient portfolios.

The following conclusions can therefore be made about domestic and foreign investments:

1. In a world of certainty the investment goal is to maximize the return. This requires an allocation of capital under which the marginal rates of return on domestic and foreign investments are equal to each other. Figure 8-2 indicates this optimal strategy. The chart shows domestic and foreign investment opportunities arranged in order of their rate of return. Given the total capital for investment $\overline{OB}$, maximum return is achieved by allocating $\overline{OA}$ to domestic investments and $\overline{AB}$ to foreign investments.
2. In the real world of uncertainty the diversification effect of a mixed investment strategy must be added to considerations of return. At any given level of expected return the proper ratio of domestic to foreign investment is that which entails the lowest total risk.

The choice of the specific level of return and the respective lowest risk is a matter of the individual country's preference. For each country the optimal

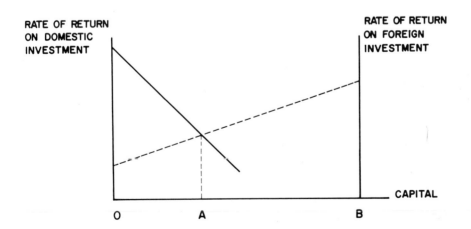

Figure 8-2. Capital Allocation for Domestic and Foreign Investment

choice is such that the ratio of incremental return to incremental risk in its investment portfolio is equal to the country's relative preference for return and risk.

We do not have to elaborate on the subject since the answer to our question is clear-cut: The most efficient strategy under either certainty or uncertainty is a mixture of domestic and foreign investments.

Standard of Living

This criterion is taken care of by the previous one—since the greater the expected return on investment at the given level of risk, the greater the future resources available to increase the standard of living. Therefore the proper policy is again a mixture of domestic and foreign investments.

National Security

Domestic investments include developing agriculture and irrigation systems; building new industries; erecting private and public housing; constructing roads, ports, communication systems, and other elements of an economic infrastructure; building schools, hospitals, and other social and public institutions; and improving the standard of other services. By strengthening the economic base, domestic investments also improve national security. Foreign investments, on the other hand, do not have a clear-cut effect on national security; in fact, since they are subject to confiscation, they may even have a negative effect on national security.

Stability of the Regime

Domestic investments have an ambivalent effect on the stability of the regime. The regime's short-run prestige is enhanced as employment opportunities are created and domestic development attains impressive proportions. In the long-run, however, the higher the rate of domestic investment, the larger will be the size of the proletariat and the intelligentsia—classes that may challenge traditional, nonmodern regimes.

The government can slow down the modernization process resulting from domestic economic development by allocating the largest share to agriculture and irrigation. However, this might produce less-than-satisfactory progress toward the other goals. Thus an effective domestic investment program cannot avoid activities that accelerate modernization, such as development of industry and expansion of education, which in the long run may make a traditional regime more vulnerable.

The Country's International Political Power

Both domestic and foreign investments strengthen a country's international political power. Again therefore, a mixture of both types of investment is the proper policy.

To summarize: domestic investment may contribute more to national security than will foreign investment. On the other hand, domestic investment may reduce the long-term power and stability of the regime. For the other three goals, the optimal investment policy is a proper mixture of both domestic and foreign investments.

In general therefore, an efficient investment strategy would divide the investment resources between domestic and foreign opportunities. The proper mixture depends on two factors: the country's priorities among the five goals and the specific investment opportunities available to the country.

A Procedure for Allocating Capital Between Domestic and
Foreign Investments

Several studies deal with capital allocation by assuming that domestic investments should be preferred to foreign ones. The only reason for allocating any capital to foreign investments would be physical constraints on domestic investment capacity. Then they explore the level of this capacity.

This procedure does not assure optimal allocation because the meaning of capacity is not well defined. One possible interpretation of capacity is the point at which the marginal return on investment falls to zero. But as long as the alternative of foreign investments provides a positive return (at the same risk level), investment in domestic projects over this point is clearly not optimal. Thus investment capacity is an ambiguous term that cannot serve as a basis for a sound policy.

We now present an alternative method for allocating capital, which we use in the following chapters. The principle is optimal allocation, i.e., an allocation that attains maximum return on the total capital. The basic condition for such maximization is equality between the marginal rates of expected returns on domestic and foreign investments at the same risk level. We use this normative optimal allocation as a basis for predicting domestic and foreign investments; thus we again apply the concept of normative prediction.

We first assume that for many OPEC members domestic investment opportunities that carry reasonably high returns are limited in relation to their huge oil revenues. This is particularly true of Saudi Arabia and the Persian Gulf Emirates (the *S* countries). The factors that constrain the attractive domestic investment opportunities in these countries are:

1. The total labor force is small, especially in Saudi Arabia and the other Persian Gulf Emirates.

2. The specialized labor force (trained workers and professionals) is severely limited. This constraint can be eased by investment in training, but this takes time.
3. The number of experienced local managers is small. Much time is required to ease this constraint.
4. Domestic markets are small and access to international markets is limited.
5. The infrastructure of roads, communications, schools, hospitals, ports, etc. is underdeveloped. Again, its development requires time.

Given the large funds available for investment and the above constraints on high-return domestic investments, it follows that foreign investments should reach a great magnitude in order to retain the basic condition of optimal allocation. Empirical studies have indicated that the overall return and risk of particularly large investment funds do not significantly differ from the average respective performance of the total market portfolio, i.e., a hypothetical portfolio consisting of all the securities traded on the major world stock markets.[1] The average annual return on such a portfolio is estimated at roughly 8 percent per year. Accordingly, we estimate OPEC members' long-term expected return on foreign investments at about 8 percent annually. Therefore domestic investments, if they are well diversified, should be determined so as to provide a marginal rate of return at about 8 percent per year.

In the next chapter we apply a model of economic growth to the Middle East OPEC members; it specifically includes, among other factors, their domestic investments. The model will determine the amount of domestic investments (and the resulting rate of growth of the Gross National Product) necessary to provide an 8 percent marginal rate of return, with foreign investments being treated as a residual carrying the same marginal rate of return. This seems to be a reasonable approximation of optimal allocation of domestic and foreign investment.

Note

1. See, for example, Sharpe W.F., "Mutual Fund Performance," *Journal of Business,* 1966, pp. 119–139; Black, Jensen, and Scholes, "The Capital Asset Pricing Model: Some Empirical Tests," in Jensen, M., *Studies in the Theory of Capital Markets,* New York, Praeger, 1972.

9

A Model of Economic Growth and Foreign Capital Accumulation

A model of future economic development of the Middle East OPEC members is presented in this chapter. The model is based on the *National Accounting Resources and Uses Report* and is applied to each year from 1973 to 1985. The economic resources of the Gross National Product are divided into three components:

1. Gross Domestic Product *excluding* the oil industry
2. Domestic Product of the oil industry
3. Income earned on foreign investments

The market value of the oil industry product depends on the price of oil. Thus the findings of Part I are essential to the analysis.

The economic uses are divided into four parts:

1. Public and private consumption
2. Domestic investments
3. Throwaway spending
4. Investments abroad

Domestic investment is determined so that its marginal rate of return is equal to that of the world market portfolio. Investments abroad are the total resources less the other three uses. Capital accumulated abroad is calculated as a cumulative sum of yearly foreign investments.

The Model

Resources

The three components of the economic resources will be denoted as:

GDP = Gross Domestic Product *excluding* the oil industry.
A = total product of the oil industry
rK = return on foreign investments, where K is the accumulated stock of foreign investments and r is the rate of return.

The Gross National Product (GNP) is the sum of these three resources. Thus for year t we have:

$$\text{GNP}_t = \text{GDP}_t + A_t + rK_t \tag{9.1}$$

Uses

Let: C = consumption
I = gross domestic investment
TS = throwaway spending
X = total exports, i.e., export of oil (including royalties) plus exports of all other goods and services
M = total imports

Equation 9.2 presents the composition of these uses, which is equal to the GNP:

$$\text{GNP}_t = C_t + I_t + TS_t + X_t - M_t \tag{9.2}$$

Foreign Investments

Equation 9.3 is derived directly from Equations 9.1 and 9.2:

$$\text{GDP}_t + A_t + rK_t - C_t - I_t - TS_t = X_t - M_t = NS_t \tag{9.3}$$

Equation 9.3 can be interpreted in two ways. It is the surplus of national income over domestic uses. It is also the net investment in foreign countries, i.e., the annual net increment in foreign capital.

We shall denote either interpretation by the term NS for net surplus. NS_t constitutes the net foreign investments in year t. We also define the term surplus S_t as the sum of $NS_t + TS_t$. Thus S_t is the surplus *before* throwaway spending is deducted; it is the maximum surplus that could be invested abroad if there was no throwaway spending.

Foreign Capital Accumulation

The accumulation of foreign capital over time has the following pattern:

$$K_{t+1} = K_t + NS_t \tag{9.4}$$

Net Import

Net import excluding oil (i.e., import minus non-oil export) is derived by calculating the difference between total domestic use and GDP. Denoting net import by *NM*, we have:

$$NM_t = C_t + I_t + TS_t - \text{GDP}_t \qquad (9.5)$$

NM provides an estimate of the net effect of oil revenue on non-oil international trade. Since it indicates the increase in non-oil net export from the oil-importing countries to the oil-producing ones, it constitutes a quantitative estimate for evaluting the oil revenue effect on the real economic sectors of the oil-importing countries. In contrast to more traditional forecasting methods, non-oil net import is not based on external assumptions as to its rate of growth; it is derived from the basic economic assumption of the macroeconomic model and is one of the outputs of this model.

Computer programming techniques were applied to this model in order to derive a large number of alternative paths of economic growth and foreign capital accumulation under various combinations of assumptions.

The Assumptions

Oil Prices

Many alternative oil price patterns were hypothesized in Part I. The following analysis presents five of them, starting as of 1976. In each of these alternatives, the actual prices are used for 1973-1975. All prices are expressed in terms of 1974 purchasing power. The five alternatives:

A_1 = constant price of \$4/bbl.
A_2 = constant price of \$8/bbl.
A_3 = constant price of \$11/bbl.
A_4 = initial price of \$7/bbl. in 1976 with the net price rising at 8 percent per year to \$12.50/bbl. in 1985
A_5 = one-shot price hike from \$4/bbl. to \$17.65/bbl. in 1980

These alternatives cover a wide range of possibilities—including a return to cheap oil (\$4/bbl.), a continuation of the present high price (\$11/bbl.), and other more optimal price patterns.

The Country Groups

In the following analysis the Middle East OPEC members are divided into three groups:

1. *S* Group—Saudi Arabia, Kuwait, United Arab Emirates (Abu Dabi, Dubai, Sharjah), Qatar, Bahrein, Oman
2. Iran
3. *LQ* Group—the remaining Middle East Members (Libya and Iraq)

These groups differ from one another mainly in the role of the oil industry in their economies, as shown by the indicators in Table 9-1.

The share of the oil industry in the economies of the *S* countries is outstanding because of their relatively small populations and low GDP. Iran's oil industry, on the other hand, although large in itself is much smaller than that of the *S* countries in relation to population and GDP. Libya and Iraq fall between the *S* group and Iran; they are not similar to each other but were grouped together to simplify the discussion.

Market Share by Country Group

The OPEC market shares of the groups appear in Table 9-2. They were calculated by the procedure described in Chapter 7. At low constant prices (A_1) or at the one-shot price hike (A_5), the *S* countries' market share rises while those of *LQ* and Iran decline. At a high constant price (A_3) or at a gradually rising price (A_4), the market share of each group remains quite constant. At a constant price of \$8/bbl. (A_2), the share of the *S* countries rises slightly at the expense of the other two groups.

Table 9-1
Economic Indicators of Middle East OPEC Members by Group

	Oil Revenue Per Capita 1974($)	Ratio of Oil Revenue to GDP 1974	Oil Reserves Per Capita (bbl.)
S Countries	4,700	12.5	26,500
LQ Countries	1,500	3.2	5,200
Iran	700	1.7	2,200

Table 9-2
Estimated Market Share of OPEC by Group
(Percentage)

Group/Year		(A_1) $4/bbl.	(A_2) $8/bbl.	(A_3) $11/bbl.	(A_4) $7/bbl. at 8%	(A_5) $4-$17.65/bbl.
S	1974[a]	53	53	53	53	53
	1980	64	53	53	54	63
	1985	71	57	53	55	63
LQ	1974[a]	21	21	21	21	21
	1980	16	21	21	21	16
	1985	12	19	21	20	16
Iran	1974[a]	26	26	26	26	26
	1980	20	26	26	25	21
	1985	17	24	26	25	21

[a]Actual market share.

Return on Foreign Investments

It is assumed that the average real rate of return on foreign investments will be 8 percent per year. A sensitivity analysis was also made for 4, 6, and 10 percent.

Investments and Gross Domestic Product

The GDP growth rate depends on the growth rates of labor and capital as well as on the rate of productivity increase. Assuming an aggregate Cobb-Douglas production function with constant returns to scale, we have the following growth relationship:

$$\frac{GDP_{t+1} - GDP_t}{GDP_t} = a\,\frac{I_t}{PC_t} + (1-a)\frac{\Delta L_t}{L_t} + PR_t \tag{9.6}$$

where
PC = total productive capital in the economy
I_t = domestic investment
L = total labor force
PR = rate of productivity increase
a = output elasticity of capital
$1-a$ = output elasticity of labor

Equation 9.6 states that the GDP growth rate is an average of the growth rates of capital and labor weighted by their output elasticity plus the rate of productivity increase.

Following the empirical findings of many countries, we assume output elasticities of one-third for capital and two-thirds for labor. We also assume a 4 percent annual increase in productivity. The average ratio of capital to output of the OPEC countries in 1973 is estimated at 1.8. The ratio between *incremental* investments and *incremental* output is higher and changes over time; this incremental ratio results from the structure of our growth model.

Equation 9.6 allows us to formulate the GDP growth rate as a function of the volume of domestic investments. Assuming that the population increases at a 3 percent annual rate, its contribution to the GDP increase is 2 percent per year.[a] Taking the annual rate of productivity increase (PR_t) at 4 percent, Equation 9.6 can be rewritten as follows:

$$\frac{\text{GDP}_{t+1} - \text{GDP}_t}{\text{GDP}_t} = \frac{1}{3} \frac{I_t}{PC_t} + 0.06 \tag{9.7}$$

Thus the rate of growth of GDP directly depends on the rate of growth of the productive capital (as measured by the ratio I_t/PC_t). This functional relationship is presented in Table 9-3. The table also indicates our assumptions as to the actual rate of growth of the productive capital and the GDP of the Middle East OPEC countries for the period 1974-1992. These rates are quite high, declining gradually to 9 percent as of 1990. At 9 percent both rates of growth are equal indicating a steady-state growth rate.

Table 9-3
Estimated Growth Rates of GDP and Productive Capital
(Percentage)

Period	Annual Growth Rate of Productive Capital	Annual Growth Rate of GDP
1974 - 1977	24	14
1978 - 1980	21	13
1981 - 1983	18	12
1984 - 1986	15	11
1987 - 1989	12	10
1990 +	9	9

[a]$(1 - a) \cdot \Delta L/L = 2/3 \cdot 3\% = 2\%$

Given the 1973 capital/output ratio for the whole economy (1.8), we derive the pattern of domestic investment, the incremental capital/output ratio, and the average capital/output ratio. The schematic growth pattern is shown in Table 9-4 where 1973 GDP = 100, and 1973 productive capital stocks = 180.

Table 9-4 summarizes the growth pattern of the GDP and the required domestic investments based on their functional relationships assumed in the model and specified in Table 9-3. The table implicitly includes the marginal rate of return on domestic investments. This rate of return can be explicitly derived from the Cobb-Douglas production function; it is the derivative of output with respect to productive capital and is equal to $a(GDP/PC)$. At the steady-state in 1990, this formula is equal to 7.8 percent. In 1992 for example, GDP = 806, PC = 3428, and the marginal rate of return is $1/3 \cdot (806/3428)$ (that is, 0.784). This rate is roughly equal to the rate of return on foreign investments (8 percent). Thus our assumptions regarding the parameters of the production function and the rate of productivity increase are consistent with the optimal allocation of investments between the domestic economy and foreign countries.

Consumption

Total private and public consumption was assumed to increase from its 1973 level. MPC_1, the marginal propensity to consume from the increase in GDP, was assumed to equal 0.8. MPC_2, the marginal propensity to consume from the increase in oil revenues, was assumed to equal 0.1. The low rate of MPC_2 is due to the fact that oil revenues are accumulated primarily by governments and governmental agencies and are not part of personal income. Thus the flow of consumption over time is as follows:

$$C_{t+1} = C_t + MPC_1 \left(GDP_{t+1} - GDP_t\right) + MPC_2 \left(A_{t+1} - A_t\right) \qquad (9.8)$$

The rates of MPC_1 and MPC_2 are subject to a sensitivity analysis.

Throwaway Spending

Nonproductive, conspicuous spending is assumed to depend on the amount of revenues from oil. The higher the current revenues, the less a government is able to resist pressures to spend some of the easy-come money for easy-go purposes. Moreover, the government itself may be tempted to initiate monumental projects which add very little to the GDP but presumably add to the government's prestige. A major throwaway item is military expenditure. We assume, therefore, that a certain proportion of oil revenue will leak in this manner.

Table 9–4
Schematic Growth of GDP and Productive Capital, 1973–1992
(1973 GDP = 100)

Year	GDP	Productive Capital Stock	Investment	GDP Increase	Incremental Capital/Output Ratio	Average Capital/Output Ratio	Investment/GDP	GDP Increase/GDP	Investment/PC
1973	100	180	43	14	2.71	1.80	0.43	0.14	0.24
1974	114	223	54	16	3.38	1.96	0.47	0.14	0.24
1975	130	277	66	18	3.67	2.13	0.51	0.14	0.24
1976	148	343	83	21	3.95	2.31	0.56	0.14	0.24
1977	169	426	89	22	4.05	2.52	0.53	0.14	0.24
1978	191	515	108	25	4.32	2.70	0.57	0.13	0.21
1979	216	623	131	28	4.68	2.88	0.61	0.13	0.21
1980	244	754	136	29	4.69	3.09	0.56	0.13	0.21
1982	306	1050	189	36	5.25	3.43	0.62	0.12	0.18
1985	421	1638	246	47	5.23	3.89	0.58	0.11	0.15
1988	566	2363	284	57	4.98	4.17	0.51	0.10	0.12
1990	679	2885	260	61	4.26	4.25	0.38	0.09	0.09
1991	740	3145	283	66	4.25	4.25	0.38	0.09	0.09
1992	806	3428	309	73	4.25	4.25	0.38	0.09	0.09

This proportion is denoted by s, so that:

$$TS_t = sA_t \tag{9.9}$$

In the following empirical analysis we assume $s = 0.20$ for the S group and 0.10 for other countries. We apply to these coefficients a sensitivity analysis.

The Output of the Model

The model described above provides alternative projections under alternative sets of assumptions in regard to the following variables.

1. Accumulation of foreign capital
2. Growth of GDP
3. Growth of GNP
4. Growth of domestic consumption
5. Growth of domestic investments and productive capital
6. Growth of net import

Some of the findings of this model are presented in the following chapter.

10

Projections of Economic Growth and Foreign Capital Accumulation

Introduction

We now summarize the main findings of our analysis of the model of economic growth described in the previous chapter. The summary presents the development of the GDP and GNP followed by a description of consumption and domestic investment flows. Then the projected net imports is derived as well as the balance of payments surplus and foreign capital accumulation. We present the findings under the assumption of a constant oil price of $11/bbl. and then summarize these findings under five alternative price patterns yielding five different sets of oil revenues. Finally, we summarize a sensitivity analysis of other assumptions underlying the model. All projections are at 1974 dollars.

Projections for $11/bbl. Constant Price Policy

This section summarizes the main outcome of the model for one alternative price pattern—a constant $11/bbl. This serves as an example of the model output and can be applied for each set of assumptions.

Sources of Gross National Product

The Gross National Product is composed of three main sources: Gross Domestic Product excluding the oil industry; net oil revenues; and return on foreign investments. Table 10-1 presents the development of these components for all Middle East OPEC members for each year from 1973 to 1980 and for 1985. Total GNP rises 466 percent from $47 billion in 1973 to about $220 billion in 1985, an annual rate of 17.5 percent. Per-capita GNP rises 320 percent from $900 in 1973 to $2900 in 1985, an annual rate of 10 percent.

In the mid 1970s net oil revenues constitute the main source of the GNP (60 percent in 1975). However, while the GDP and the return on foreign investments show a trend of increase, the net oil revenues gradually decline. By 1980 they constitute only 36 percent of the GNP, and by 1985, less than 20 percent.

91

Table 10-1
Estimated Sources of Gross National Product, 1973-1985
(Billions of Dollars)

Year	Gross Domestic Product	Net Oil Revenues	Return on Foreign Investments	Total GNP
1973	33	13	1	47
1974	38	72	2	112
1975	42	72	6	120
1976	48	70	10	128
1977	56	68	13	137
1978	64	66	17	147
1979	72	64	21	157
1980	82	60	24	166
1985	143	40	36	219

Domestic Uses of Gross National Product

The domestic uses of the GNP include private and public consumption, domestic investment, and throwaway spending. Consumption rises with the GDP; the marginal propensity to consume is 0.8. Consumption is also affected by the level of oil revenues since 10 percent of the revenue is assumed to be allocated to consumption. Total consumption rises about 450 percent between 1973 and 1985; per-capita consumption rises 320 percent from $480 in 1973 to $1540 in 1985, an annual rate of 10 percent. See Table 10-2.

Domestic investment rises gradually, as required for GDP growth; it reaches its peak in 1983 ($77 billion) and then tapers off slightly. The main reason for the post-1983 decline in domestic investment lies in Iran's lack of sufficient resources in the 1980s for continued intensive investment (see below). The ratio of domestic investment to GNP is 18 percent in 1975, 32 percent in 1980, and 30 percent in 1985. The respective ratios to total domestic spending are 30 percent, 40 percent, and 35 percent.

Throwaway spending is assumed to constitute 20 percent of the revenues in the S group and 10 percent in the other groups; however, these are mere gussees. Throwaway spending does not affect the estimate of consumption, domestic investment, and the GDP; it affects only the accumulation of foreign capital and net imports.

Net Import

Net import is the difference between total domestic spending and GDP. It measures the surplus of total import over non-oil export. Net import increases

Table 10-2
Estimated Domestic Uses of Gross National Product, 1973-1985
(Billions of Dollars)

Year	Consumption	Domestic Investments	Throwaway Spending	Total Domestic Spending
1973	26	14	2	42
1974	35	17	11	63
1975	39	22	11	72
1976	44	27	10	81
1977	50	33	10	93
1978	56	36	10	102
1979	62	44	10	116
1980	69	53	9	131
1985	116	66	6	188

rapidly in 1974 and rises gradually from 1975 onward (see Table 10-3). The 1974 jump expresses the adjustment of the economies toward a greater rate of economic growth and a higher standard of living as a result of the oil price hike.

Foreign Investments and Capital Accumulation

Foreign investments are the difference between GNP and domestic spending. Foreign capital accumulation is the accumulated sum of foreign investments (see Table 10-3). The annual investment abroad (the net surplus in the current account of the balance of payments) jumps upward after the 1974 oil price hike. A few years at a constant level follow, with a slight declining trend that accelerates in 1979. After adjusting to the lower level of the Iranian economy in the early 1980s, foreign investment stabilizes at $30 billion/year.

Capital accumulation increase accordingly, reaching $290 billion in 1980 and $440 billion in 1985.

Projections for OPEC Groups

The above economic indicators can be disaggregated for each of the three groups—Saudi Arabia and the Persian Gulf Emirates (*S*), Iran, and Libya and Iraq (*LQ*).

The *S* Group. Table 10-4 summarizes the main projections for the *S* group. Since the capital available for economic growth is much greater than required, the development of each factor increases smoothly and consistently. The net

Table 10-3
Estimated Net Import, Foreign Investments, and Capital Accumulation,
1973-1985
(Billions of Dollars)

Year	(1) GDP	(2) Total Domestic Spending	(3) GNP	(4) = (2) – (1) Net Import	(5) = (3) – (2) Foreign Investments	(6) = Σ (5) Foreign Capital Accumulation
1973	33	42	47	9	5	10
1974	37	63	112	26	49	15
1975	42	72	120	30	48	64
1976	48	81	128	33	47	112
1977	56	93	137	37	44	159
1978	64	102	147	38	45	203
1979	72	116	157	44	41	248
1980	82	131	166	49	35	289
1985	143	188	219	45	31	440
Total Increase (%)	433	448	466			
Annual Increase (%)	13.0	13.3	13.7			

surplus for foreign investment is constant at approximately $30 billion/year. Total S-group accumulation reaches almost $200 billion in 1980 and $350 billion in 1985, constituting 65 percent of total Middle East OPEC accumulation in 1980 and 78 percent in 1985.

Iran. Iran's economic development trends are not as smooth as those of the S countries. Its net surplus for foreign investment reaches $10 billion in 1974, then gradually falls—becoming negative in 1980. Consequently, its foreign capital accumulation, which rises slowly in the 1970s, reaches its peak near 1980 and declines during the 1980s. This limits domestic investment, which decreases from 1984, and the GDP growth rate slows down accordingly. See Table 10-5.

From the projections in Table 10-5 it appears that Iran's economy, which is bound to achieve a high level of development, will continue to grow at somewhat lower rates after 1985 when the oil industry's share in the economy declines to a low level.

Libya and Iraq. GDP growth in Libya and Iraq (LQ) is rapid and undisturbed. However, net surplus for foreign investment declines—reaching zero by 1985 and probably negative levels after that. See Table 10-6.

Table 10-4
Estimated Main Economic Development Indicators for the *S* Countries,
1973-1985
(Billions of Dollars)

Year	(1) GDP	(2) Total Domestic Spending	(3) GNP	(4) = (2) - (1) Net Import	(5) = (3) - (2) Foreign Investments	(6) = Σ (5) Foreign Capital Accumulation
1973	8	11	16	3	5	6
1974	9	21	49	12	28	11
1975	11	34	52	13	28	39
1976	12	26	55	14	29	87
1977	14	28	58	14	30	96
1978	16	30	61	14	31	126
1979	18	34	65	16	31	157
1980	20	37	68	17	31	188
1985	36	55	85	19	30	350
Total Increase (%)	450	500	531			
Annual Increase (%)	13.3	14.4	15.0			

The *LQ* group constitutes a middle ground between the *S* Group and Iran. Net capital accumulation rises slowly and declines after 1985; domestic spending increases gradually.

Comparison of Groups

Gross Domestic Product increases at similar rates in each group but Gross National Product does not. Excluding the oil revenue jump in 1973-1974, the *S* countries' annual growth rate from 1975 to 1985 is smaller than that of Iran (5 percent versus 7 percent) because oil revenues constitute a greater percentage of GNP in the *S* countries; since these revenues do not grow, they reduce the GNP growth rate. On the other hand, per-capita GNP is much greater in the *S* group than in Iran simply because per-capita oil revenues are significantly greater in the *S* group.

The highest per-capita consumption is in the *S* group where it reaches almost $2,300 in 1985. The lowest is in Iran ($1,300 in 1985). This difference results from the difference in per-capita GNP.

Table 10-5
Estimated Main Economic Development Indicators for Iran, 1973-1985
(Billions of Dollars)

Year	(1) GDP	(2) Total Domestic Spending	(3) GNP	(4) = (2) - (1) Net Import	(5) = (3) - (2) Foreign Investments	(6) = Σ (5) Foreign Capital Accumulation
1973	17	21	20	4	−1	2
1974	19	28	38	9	10	1
1975	21	32	41	11	9	11
1976	24	36	44	12	8	20
1977	28	43	48	15	5	28
1978	32	47	52	15	5	33
1979	36	54	56	18	2	38
1980	41	63	60	22	−3	40
1985	71	81	82	10	1	10
Total Increase (%)	418	386	410			
Annual Increase (%)	12.7	12.0	12.5			

Because of common assumptions regarding the aggregate production function and GDP growth, domestic investments behave similarly in all three groups. In the 1980s however, Iran's investment declines because of insufficient resources, while the other countries retain their high level of investment.

The S countries' net import in 1975 accounts for over 45 percent of the total, while that of Iran amounts to about 35 percent. Iran's share in net import rises more rapidly, reaching 42 percent in 1980 (versus 34 percent for the S group). After 1983 the decline in Iran's investments lowers its net imports, carrying the total downwards; furthermore, Iran's share of the total declines substantially.

The S countries' foreign capital accumulation amounts to over 60 percent of the total in the 1970s and rises to 78 percent by 1985. This increase comes at the expense of Iran, whose share falls from 18 percent in the 1970s to zero after 1985. This, too, is attributed to Iran's lack of resources for domestic investment in the 1980s.

Projections at Five Alternative Oil Prices

Table 10-6
Estimated Main Economic Development Indicators for the *LQ* Countries, 1973-1985
(Billions of Dollars)

Year	*(1)* GDP	*(2)* Total Domestic Spending	*(3)* GNP	*(4) = (2) - (1)* Net Import	*(5) = (3) - (2)* Foreign Investments	*(6) = Σ (5)* Foreign Capital Accumulation
1973	8	10	11	2	1	2
1974	9	14	25	5	11	3
1975	10	16	27	6	11	14
1976	12	19	29	7	10	25
1977	14	22	31	8	9	35
1978	16	25	34	9	9	44
1979	18	28	36	10	8	53
1980	21	31	38	10	7	61
1985	36	52	52	16	0	80
Total Increase (%)	450	520	473			
Annual Increase (%)	13.3	14.8	13.8			

Oil Revenues

The net oil revenues expected to result from the five alternative price patterns appear in Table 10-7. These revenues are derived from the model developed in Part I.

Not all alternatives are reasonable predictions (e.g., A_1), but they are presented here in order to provide a wide spectrum for comparison. No alternative presents a price greater than $11/bbl. in the 1970s; at 10 percent annual inflation, however, $11/bbl. at 1974 prices is equivalent to $19.50/bbl. in 1980. Moreover, at a higher real price, oil revenues in real terms decline significantly.

Gross National Product

Gross National Product rises at 13 to 15 percent per year. The main component of GNP—the Gross Domestic Product, excluding the oil industry—is not affected by the oil price level; GDP rises at 13 percent annually at all price levels. The oil

Table 10-7
Estimated Net Oil Revenues for Middle East OPEC Members at Alternative Price Patterns, 1974-1985
(Billions of Dollars at 1974 Prices)

Year	(A_1) $4/bbl.	(A_2) $8/bbl.	(A_3) $11/bbl.	(A_4) $7/bbl. at 8%	(A_5) $4-$17.65/bbl.
1974[a]	72	72	72	72	72
1975[a]	71	71	71	71	71
1976	22	50	70	43	22
1977	25	51	68	48	25
1978	27	53	66	53	27
1979	30	54	64	59	30
1980	33	56	61	64	175
1981	36	57	58	69	160
1982	39	58	54	76	142
1983	42	60	50	73	122
1984	46	61	45	79	98
1985	49	63	40	79	70

[a]The revenues for 1974 and 1975 are based on actual prices and thus are the same in all alternatives.

price does, of course, affect oil revenues and foreign capital accumulation. As a result, at low prices GNP (which includes net oil revenues and return on foreign investments) is $210 billion in 1985 while at optimal prices it reaches a higher level—over $250 billion.

Consumption and Domestic Investment

Since consumption is determined mainly by GDP and less by net oil revenues, its sensitivity to oil prices is insignificant. Thus it rises 13 to 14 percent annually for all alternatives.

When oil revenues are not high enough to finance the domestic investments required for economic growth, these investments become sensitive to the level of revenues. In the S countries, oil revenues are always large enough. In Iran, however, domestic investment sensitivity is very large—especially in the 1980s. Iran reaches a domestic investment of $40 billion in 1985 with the optimal price policy as opposed to only $18 billion at low oil prices.

Interestingly, high prices (such as $11/bbl.) yield high revenues only in the early period. In later periods, oil revenues decrease with the increasing supply of competitors. Consequently, the 1985 oil revenues at $11/bbl. are only slightly higher than those at $4/bbl. Thus Iran's 1985 investments at $11/bbl. will also be low.

Net Import

Net import excluding oil is sensitive to oil prices. It reflects the sensitivity of total domestic spending (which includes domestic investments). Net import in 1985 is about $35 billion at low prices. At optimal prices it is $70 billion. Once again, sensitivity is most striking in the case of Iran.

Foreign Investment and Capital Accumulation

Foreign investment is greatly affected by oil prices. It varies from $16 billion to $35 billion in 1980, and from $30 billion to $50 billion in 1985. Consequently, total foreign capital accumulation is also affected by oil price. It varies between $170 billion and $300 billion as of 1980, and between $270 billion and $640 billion as of 1985.[a] One should note, however, that the difference among three of the alternative price patterns ($8, $11, and $7 at 8 percent) is relatively small. See Table 10-8.

The foreign capital is concentrated mainly in the S countries; they hold about 75 percent of the total in 1985—and if prices are low, their holding reaches over 90 percent. The LQ group accounts for the remainder; Iran has no significant foreign capital holding in any case.

Table 10-8
Estimated Foreign Capital Accumulation at Alternative Price Levels, 1973-1985 (Billions of Dollars)

End of Year	(A_1) $4/bbl.	(A_2) $8/bbl.	(A_3) $11/bbl.	(A_4) $7/bbl. at 8%	(A_5) $4-$17.65/bbl.
1973	20	20	20	20	20
1975	74	74	74	74	74
1980	170	250	300	240	170
1985	270	410	440	450	640

[a]For alternative forecasts see Appendix.

Sensitivity Analysis

These projections are all based on one set of assumptions regarding the behavior of the main economic parameters. A sensitivity analysis for other parameters tested the significance of these assumptions. Some of the findings are presented below.

Marginal Propensity to Consume

Consumption is affected by the marginal rate of consumption from income; the lower the rate, the smaller is the consumption level. This reduces net import to a small extent. It does not affect investments—except in Iran, where it slows down the reduction of investment.

GDP Growth Rate

If the GDP growth rate decreases, domestic investment is lower. Net import is therefore lower, and foreign capital accumulation is greater. For instance, a decline of 1 percent in the GDP growth rate results in an additional accumulation of over $30 billion by 1980.

Productivity Rate

A decline in the rate of productivity increase affects the real GDP figures, but its impact on the accumulation of capital and net import is insignificant. This is true because a decline in productivity has no effect on investment and only a slight effect on consumption. If domestic investment is increased to make up for lower productivity (thus maintaining the GDP growth rate), smaller accumulation of capital results.

Rate of Return on Foreign Investments

On the basis of our assumptions, a change in the rate of return on foreign investment has no effect on domestic spending and, therefore, has no effect on net imports either. Such a change would affect Iran's investments in the years that they are limited by insufficient oil revenues. The main effect of a change in the rate of return is felt on the rate and level of accumulation of foreign capital. For example, if the rate of return falls from 8 percent to 4 percent, total 1980 accumulation at \$11/bbl. falls about 10 percent from \$290 billion to \$260 billion; in 1985 the respective accumulations are \$450 billion and \$340 billion— a difference of 25 percent.

Shifting Revenues among Countries

The above findings are based on a given procedure for determination of each group's share of OPEC sales. If one group's output of oil increases more than another group's does, its share of the revenue will increase. How does this affect the results? First, it changes the ownership distribution of foreign capital. However, if lower revenues are assumed, it has more important effects. Consider a case in which Iran is unable to reach its highest domestic investment plans. If revenues are shifted from the other countries (especially the S Group) to Iran, the other countries' foreign capital accumulations will decline accordingly. Iran, on the other hand, will not accumulate this capital but will invest it domestically. This will increase Iran's GDP growth rate without reducing the rates of the other countries. There will be an increase in total net import and an equivalent decrease in total capital accumulation.

The relative size of this effect was analyzed. It was found that if a certain sum of revenue was shifted from the S group to Iran, and if this shift was distributed over the period in constant annual amounts, then the accumulation of capital would be smaller by 85 percent of the revenue shifted. For example, if \$1 billion is shifted from the S group to Iran each year from 1974 to 1985 (a total of \$12 billion), the accumulation of foreign capital by 1985 will be smaller by \$10 billion. This holds until Iran reaches its optimal level of domestic investment.

Part 3

Investment Policy for OPEC Foreign Capital

11 Investment Policy—Principles

Introduction

Estimates of foreign capital accumulation by the Middle East OPEC members were presented in Chapter 10. The estimates for 1980 vary (with oil prices) from $170 billion to $300 billion (in 1974 prices). We now discuss the problem of an investment policy for this capital and present a methodological framework by which *optimal* investments can be made. Given an investor's preferences, and assuming that investors behave rationally in accordance with their preferences, such a method helps to devise an investment program for maximum achievement of investment goals. This program is, in fact, an investment portfolio—a set of specific investment forms in prescribed proportions. Such an investment program can be interpreted in a positive sense; given the investor's preferences and the investment opportunities, the optimal portfolio might be regarded as the forecasted one. Again, this is normative forecasting—as in the case of optimal price determination discussed in previous chapters.

In this chapter we briefly review the investment criteria and the different forms of investment available to the capital owners. In the next chapter we present an applied model for optimal investment.

Investment Criteria

Investors have explicit and implicit goals which they try to fulfill by chosing a proper investment portfolio. A preliminary set of goals was presented in Chapter 8:

1. To increase the expected rate of return
2. To reduce business risk
3. To reduce political risk
4. To increase national security
5. To increase the country's international political power
6. To increase the stability of the regime.

In addition, those investments that are more easily manageable are preferable; the lack of managerial resources constitutes an effective constraint on investment strategy.

105

These goals are stated in an ordinal form, since most of them are not measurable in a cardinal sense. Furthermore, the existence of multiple goals makes it impossible to devise an investment portfolio that is superior on every criterion. For example, an investment set which has a greater expected rate of return than another set may be politically riskier; a portfolio that contributes more to national security than another portfolio may have a greater business risk or a smaller expected rate of return.

Any investment portfolio will achieve a certain level of each goal. If the levels of each goal achieved by one portfolio are less than the respective levels achieved by another set of investments, then the first portfolio is clearly inefficient. We must discard all such inefficient portfolios and retain the efficient ones. An efficient portfolio is one that, for any given level of achievement of any $n - 1$ criteria, has the highest level of achievement of the remaining criterion. This definition of efficient investment is objective since it is independent of the investor's subjective preferences for certain criteria.

Different efficient investment sets can be defined for each combination of goals. Consequently, the number of such sets is virtually infinite. For a certain investor, however, there might be one efficient set which provides him with the greatest level of his *subjective* preferences; this is his optimal investment portfolio. Thus different investors will choose different optimal investment portfolios—depending on their subjective goal preferences.

Investment Forms

Following are the general classes of investment forms available to the oil-producing countries:

1. Short-term financial investments in the money markets of the developed countries.
2. Long-term financial investments in the securities markets of the developed countries. These investments are classified into two main forms:

    a. Commercial and institutional bonds
    b. Corporate stocks of various types purchased on the basis of their financial performance, with no direct contact with the corporations themselves. This group will be termed *stock portfolio.*

3. Direct investments in industrially developed countries. A direct investment is distinct from a financial investment or a portfolio in that it implies that the investor can exercise some degree of control over the corporation whose shares are held. We classify these investments into three groups:

    a.   Minority control, where the investor owns a relatively small proportion of the corporate stock; some influence can be exercised, but not enough to affect corporate policy.

    b.   Majority control, where the investor owns a substantial proportion of the corporate stock; here the investor can have a substantial influence on management appointments and corporate policy.

    c.   Joint ventures with a corporation in the corporation's country.

4.   Neutral investments in developed and underdeveloped countries. These are direct investments in which it makes little difference to the country whether the ownership is foreign or domestic (such as real estate, hotels, shopping centers).

5.   Investments in the energy industry anywhere in the world.

6.   Investments in neighboring and underdeveloped countries.

The formulation and implementation of an investment strategy requires time. During this period, the capital must be held in some temporary form that does not prevent its later investment according to the strategy. The natural form of temporary investment is in bank deposits and short-term money market instruments. These short-term investments also provide liquidity for the conduct of foreign trade and for long-term investment transactions. Therefore, even after the long-term investment strategy is implemented, short-term investments will continue to be held—although probably in a smaller proportion.

12 Investment Policy—Application

This chapter presents a technique for optimizing investment policy and a specific application of the technique. The application is based on tentative data and assumptions; therefore it should be regarded as indicative—serving as an example of the methodology rather than as a realistic policy.

We first present a set of coefficients of investment performance, roughly indicating the contribution of each investment form to each criterion. Next, we develop a quadratic programming model that allocates capital to the different investment forms so as to maximize the sum total of the (weighted) investment criteria. We then apply this model to our specific problem and solve the model on tentative data.

Investment Performance

It is difficult to establish a meaningful quantitative relationship that would show the contribution of a $1 investment in a certain form toward achievement of each goal. Quantitative empirical estimates may be derived only on the first two criteria—expected rate of return and business risk—and even here the data is incomplete. The remaining criteria are not subject to objective measurement and therefore cannot be quantified meaningfully.

Consequently, we developed an experimental matrix of coefficients graded from 1 to 5, roughly indicating the contribution of each investment form toward each goal. These coefficients are only tentative since they are based on intuition rather than on a broad, formal survey. In the matrix we have classified the investment forms into nine main groups. See Table 12-1. In the table, the higher the coefficient, the greater is the contribution of a $1 investment toward achievement of the criterion. Thus investments associated with higher return have higher coefficients; those associated with greater risk have smaller coefficients.

The right-hand column indicates the relative input of managerial resources required to implement $1 of investments. Due to the scarcity of managerial resources, this input constitutes a constraint on the investment program. The higher the coefficient in this column, the more managerial resources are required.

The table indicates that short-term assets are inferior to bonds except in regard to business risk. They are also inferior to most other forms of investment

Table 12-1
Tentative Coefficients of Investment Performance

	Maximizing Return	Minimizing Business Risk	Minimizing Political Risk	Increasing National Security	Increasing Political Power	Increasing Stability of Regime	Input of Managerial Resources
Short-term Assets	2	5	5	1	1	1	1
Bonds	3	4	5	2	5	2	1
Stock-Portfolio	4	3	4	1	1	1	4
Stock Minority	4	2	2	3	3	1	3
Stock Majority	3	1	1	3	4	1	5
Joint Venture	4	2	2	3	4	1	4
Neutral Investment	5	1	3	1	1	1	4
Energy	4	1	2	1	5	3	5
Underdeveloped Countries	1	1	4	3	5	4	1

except in regard to business and political risk. However, as mentioned above, short-term assets are held for two other reasons (to provide liquidity for foreign trade and for long-term investment transactions) that are not applicable to other investment forms. Therefore we do not use the coefficients to determine the proportion of short-term assets in the portfolio; instead we assume that their proportion is determined first by other considerations. Then the shares of the remaining investment forms are solved.

Investments in neighboring and underdeveloped countries are expected to provide a negative rate of return. For this reason, they are also excluded from the following exercise; we assume that their share is determined exogeneously.

The Model

This section presents a model that determines an optimal portfolio based on the preceding matrix of investment performance. The optimal portfolio is a set of investment forms in a proportion that achieves the maximum total value of performance indicators. However, these indicators should be summed up only after being weighted by the relative importance (or preference) of the investment criteria.

Let i,j = investment form
k = criterion
X_i = proportion of investment form i in the portfolio
E_{ik} = value of the performance coefficient of investment i at criterion k
a_k = preference coefficient of criterion k
m_i = coefficient of managerial input at investment form i
M = total managerial resources

The investment criteria are now classified into two groups: criteria of risk (reducing business and political risk) and criteria associated with return. The combined business and political risk of any investment form is measured by the variance of its return. The risk of the total portfolio is also measured by its variance, which is the total of the variances and covariances of its components weighted by their proportions in the portfolio. Thus:

$$\sigma_p^2 = \sum_i \sum_j X_i X_j \, \text{cov}(ij)$$

$$= \sum_i \sum_j X_i X_j \rho_{ij} \sigma_i \sigma_j$$

(12.1)

where σ_p^2 = variance of the portfolio return
σ_i, σ_j = standard deviation of the returns on investment form i, j
$\text{cov}(ij)$ = covariance of the returns of investment forms i and j
ρ_{ij} = correlation coefficient between the returns of investment forms i and j

The model that maximizes the sum total of the weighted criteria can now be formulated as quadrative programming:

$$\max \sum_i \sum_k \chi_i a_k E_{ik} - a_2 \sum_i \sum_j \chi_i \chi_j \rho_{ij} \sigma_i \sigma_j \qquad (12.2)$$
$$k \neq 2,3$$

subject to:

$$\sum_i \chi_i m_i \leqslant M$$

$$\sum_i \chi_i \leqslant 1$$

$$\chi_i \geqslant 0$$

where $k = 2,3$ refers to business risk and political risk and $a_2 = a_3$

Application

This quadrative programming model was applied to the performance matrix of Table 12-1, where the investment forms are denoted as follows:

1 = bonds
2 = stock portfolio
3 = stock minority
4 = stock majority
5 = joint venture
6 = neutral investment
7 = energy

The following additional assumptions were made:

1. The expected return and variance of the seven investment forms were assumed as follows:

Investment Form	1	2	3	4	5	6	7
Expected Rate of Return ($\overline{r_i}$)	6	10	10	8	10	12	10
Variance (σ_i^2)	1	11	15	18	16	19	21

2. The correlation coefficient between all pairs of investments was assumed to be 0 except for the following pairs:

	X_5	X_6	X_7
X_5	1	1/4	1/2
X_6	1/4	1	1/4
X_7	1/2	1/4	1

3. The investment performance of the remaining criteria associated with return (national security, political power, stability of regime) are measured by the coefficients shown in Table 12–1. In the model they are transformed into "equivalent expected rate of return" by a factor of 0.5 applied to each criterion. Thus, $a_4 = a_5 = a_6 = 0.5$
4. The preference coefficients for the risk factors (a_2, a_3) are assumed to equal -0.25. The sign is negative since the greater the risk, the smaller the level of satisfaction provided by the investment portfolio.

Given these assumptions, the model becomes:

$$\max Z = \sum_{i=1}^{7} x_i \bar{r}_i + 0.5 \sum_{i=1}^{7} \sum_{k=4}^{6} x_i E_{ik} - 0.25 \sum_{i=1}^{7} \sum_{j=1}^{7} x_i x_j \rho_{ij} \sigma_i \sigma_j \quad (12.3)$$

subject to:

$$\sum_{i=1}^{7} x_i m_i \leqslant M$$

$$\sum_{i=1}^{7} x_i \leqslant 1$$

$$x_i \geqslant 0$$

The objective function Z can now be interpreted as the "certainty equivalent rate of return" on the investment portfolio. The first component is simply the expected rate of return. The second component is the equivalent expected rate of return of the nontangible criteria. The third component is the risk premium, i.e., risk evaluated by its substitution coefficient to expected return. This risk premium is deducted from the first two components to reach the certainty equivalent rate of return.

Solution

The model was solved under the asumption that the management constraint M is 3. The "optimal" proportions of the investment forms appear in the Table 12-2, Solution One column.

The value of the objective function in Solution One (12.21) is interpreted as a percentage showing the certainty equivalent rate of return on the investment portfolio. However, this should not be compared with empirical market rates of return because our measure includes equivalent returns of intangible criteria (national security, political power, and stability of regime).

Under Solution One all investment forms enter the optimal portfolio. Bonds take the greater share—40 percent. Stock minority and energy are second with 18 percent and 17 percent. The other four forms take small proportion (between 3 and 8 percent each), with stock majority having the smallest share.

This solution is indeed artificial. It results from a combination of assumptions—the most crucial assumptions being the expected rates of return, the variances and covariances, the performance matrix of the remaining criteria, the weights associated with the investment criteria, and the input coefficient and overall constraint of managerial resources.

For this reason, the solution should not necessarily be interpreted as a recommended investment policy. It should serve as a basis for investment policy only if the above assumptions are assumed to be correct. In such a case, it could also serve as a basis for normative forecasting of the future investment behavior of petro-money owners.

Table 12-2
Optimal Portfolios under Alternative Assumptions

		Solution One	Increased Cost of Risk	Reduced Managerial Resources
1.	Bonds	0.40	0.56	0.70
2.	Stock portfolio	0.06	0.07	
3.	Stock minority	0.18	0.12	0.10
4.	Stock majority	0.03	0.04	
5.	Joint venture	0.08	0.06	0.05
6.	Neutral investment	0.08	0.06	0.04
7.	Energy	0.17	0.09	0.11
	Total	1.00	1.00	1.00
	Objective Function (%)	12.21	11.50	11.93

Although the above assumptions are arbitrary, the model can nevertheless be used to test the effects of these assumptions on the optimal investment portfolio. This can be done by a sensitivity analysis—changing assumptions and comparing the differences among the resulting solutions.

We made a great number of sensitivity analyses. Table 12-2 presents two of them:

1.   Increased cost of risk, from $a_2 = -0.25$ to $a_2 = -0.50$

The higher the cost of risk, the greater is the proportion that will be invested in lower-risk forms (bonds) and the smaller is the proportion that will be invested in higher risk forms (energy and stock minority). Again, the specific magnitude of the effect depends on the other assumptions as well—particularly the variances and covariances of the investment returns. The value of the objective function is lower (11.50 percent compared with 12.21 percent in Solution One) simply because risk is evaluated by a greater cost coefficient.

2.   Reduced managerial resources, from $M = 3$ to $M = 2$.

The smaller the managerial resources, the greater will be the proportion of investment forms that require little managerial input and the smaller will be the proportion of forms involving greater managerial input. The value of the objective function is again lower than in Solution One (11.93 percent versus 12.21 percent). This means that the managerial constraint is effective, depriving the investors from making more attractive investments that require greater managerial input than is available. The difference in the value of the goal function measures the marginal value of the management input resulting from a reduction in the available resources by one unit. This is the shadow price of the management constraint.

Investments with greater risk naturally involve greater managerial input—so the effect of reduced managerial resources is similar to that of increased cost of risk: the proportion of bonds rises, and the shares of energy and stock minority decline. In our exercise, stock portfolio and stock majority are virtually excluded when managerial resources are reduced.

Appendix

Projections of OPEC Oil Output, Revenue, and Capital Accumulation

Since the sharp increase in the price of oil at the end of 1973, many projections have been made of OPEC oil output, revenues, and foreign capital accumulation. These projections differ from one another because they are based on different sets of assumptions. The main assumptions affecting the projections refer to the future level of oil prices, the effect of the price on demand for energy and for OPEC oil, the rate of increase in OPEC imports, and the world inflation rate.

This Appendix compares a number of these projections with estimates derived from our study. Our estimates are made here by applying the explicit assumptions of the comparable projections to the two models presented in Chapters 5 and 9.

Our estimates are compared with the following projections:

1. Future OPEC Accumulation of Oil Money: A New Look at a Critical Problem," Walter Levy Consultant Corporation, New York, June, 1975 (hereafter referred to as Levy).
2. H.B.Chenery, "Reconstructing the World Economy," *Foreign Affairs,* January, 1975, pp. 242-263 (hereafter referred to as Chenery).
3. "Why OPEC's Rocket will Loose its Thrust," *First National City Bank Monthly Review,* June, 1975 (hereafter referred to as FNCB).
4. Arnold E. Safer, "Outlook for World Oil: Prices and Petrodollars," *Economic View from One Wall Street,* Irving Trust Company, March 20, 1975 (hereafter referred to as Irving).
5. *Prospects for the Developing Countries,* International Bank of Reconstruction and Development, July 8, 1974 (hereafter referred to as World Bank).
6. "Oil, Looking Back and Looking Ahead," *World Financial Markets,* Morgan Guaranty Trust Company of New York, January 21, 1975 (hereafter referred to as Morgan).

The Appendix is divided into three sections: (1) Oil Output, (2) Oil Revenues, and (3) Foreign Capital Accumulation.

Projections of OPEC Oil Output

Levy's Projection

Table A-1 presents Levy's projection of OPEC oil output and prices for 1977 and 1980. By inserting Levy's prices (in 1974 dollars) into our model, we derive our respective output estimates.

The two sets of projections are close to each other. One should note, however, that our estimates change with the inserted prices. At other price vectors the estimates will, of course, be different.

Chenery's Projections

Chenery's projections for OPEC output for 1980, 1985, and 1990 are based on estimates of OECD and the World Bank. Table A-2 compares Chenery's estimates with ours.

Output projections at $7/bbl. are close to each other, even as far into the future as 1990. The estimates at $9.60/bbl. are close for 1980 and 1985 but not for 1990.

FNCB Projections

The First National City Bank's output projections are compared to ours in Table A-3. Although the two projections *seem* to be close to each other for the period preceding 1980, they do differ significantly. In the FNCB projection, the effect of a real price decline on the increase in total demand for OPEC oil is very slight. This is shown when the projection is extended to 1985. At a low price range ($5/bbl. to $6/bbl. in 1974 dollars) in the 1980s, FNCB projects an increase of less than 4 percent in OPEC output—from 11.3 bil. bbl. in 1980 to 11.7 bil. bbl. in 1985. Our projection at the same price is for a much greater increase—36 percent.

Irving Projections

Irving Trust's output projections from 1977 onward appear to be about 20 percent lower than our estimates. This is shown in Table A-4. A second projection presented by Irving is based on an assumption that prices will begin to break in 1977, declining gradually to $4.50/bbl. as of 1980 (in 1974 prices). However, the projected output is left unchanged. This contradicts our basic premise that the price level affects the demand for OPEC oil.

Table A-1
Comparison with Levy's Projections of OPEC Oil Output, 1977-1980

	1977	1980
Current Price ($/bbl.)	12.00	14.65
Real Price (1974 = 100)[a] ($/bbl.)	9.40	9.30
Levy's Estimates (bil. bbl.)	11.50	11.50
Author's Estimates (bil. bbl.)	11.20	11.90

[a]Projected price index: 1977 = 128; 1980 = 157.

Table A-2
Comparison with Chenery's Projections of OPEC Oil Output, 1980-1990 (Bil. Bbl.)

	Constant Price $9.60/bbl.		Constant Price $7/bbl.	
	Chenery's Estimates	Author's Estimates	Chenery's Estimates	Author's Estimates
1980	12.0	11.8	15.3	14.1
1985	13.1	12.5	17.9	17.4
1990	9.5	13.1	20.1	21.5

Table A-3
Comparison with FNCB's Projections of OPEC Oil Output, 1975-1985

Year	Current Price ($/bbl.)	1974 Price[a] ($/bbl.)	FNCB Estimates (bil. bbl.)	Author's Estimates (bil. bbl.)
1975	11.30	10.10	9.5	10.7
1976	11.80	10.00	9.9	10.9
1977	11.20	9.10	10.2	11.1
1978	10.70	8.20	10.6	11.5
1979	9.90	7.20	11.0	12.1
1980	9.10	6.30	11.3	12.8
1985	9.10	5.00	11.7	17.4

[a]The FNCB price at 1975 dollars is adjusted here to the 1974 price level by a factor of 12 percent.

Table A-4
Comparison with Irving's Projections of OPEC Oil Output, 1974-1980

Year	Current Price ($/bbl.)	Price Index[a]	1974 Price ($/bbl.)	Irving Estimates (bil. bbl.)	Author's Estimates (bil. bbl.)
1974	9.40	100	9.40	11.5	
1975	11.00	112	9.80	10.8	10.7
1976	12.00	119	10.10	10.0	10.8
1977	12.87	128	10.10	9.0	11.0
1978	13.74	137	10.00	9.0	11.2
1979	14.70	147	10.00	9.0	11.3
1980	15.72	157	10.00	9.0	11.4

[a]Based on Levy's estimates.

The World Bank Projections

The World Bank's output projections were made earlier than the others presented above. It is compared with our estimates in Table A-5. The two estimates are very close to each other at a price of $7/bbl. They differ widely, however, at $11/bbl. The World Bank estimate is considerably higher and shows an increase during the 1980s, while our estimate shows a slight decline.

Morgan Guaranty projections refer to oil revenues without specifying its assumptions on output and prices. Therefore we are unable to make a comparison of projected outputs.

Projections of OPEC Oil Revenues

Total net revenues for OPEC are inferred directly from each price pattern and its associated demand for OPEC oil. They are calculated by multiplying the projected

Table A-5
Comparison with World Bank Projections of OPEC Oil Output: 1980, 1985
(Bil. Bbl.)

Year	Constant Price $11/bbl.		Constant Price $7/bbl.	
	World Bank Estimates	Author's Estimates	World Bank Estimates	Author's Estimates
1980	12.9	10.6	14.6	14.1
1985	15.2	9.9	17.7	17.4

output of OPEC oil (from the previous section) by the net price (market price minus direct cost). The resulting net revenues are in 1974 dollars. The projections can be adjusted to current dollars by inflating these revenues at expected rates of world price increase. Since the projected revenues result directly from the projections of output and the associated prices, there is no need to repeat the process of comparison made in the preceding section. We shall, therefore, make only brief comparisons.

In Table A-6, three alternative oil revenues are presented. They are derived from our model under constant prices of $8/bbl., $10/bbl., and $12/bbl. and are adjusted to current dollars by a price index used by Levy. The table shows that OPEC revenues at 1980, after adjustment for inflation, are expected to reach $130 billion to $160 billion. These estimates are based on a 1980 oil price (in current prices) of $13/bbl. to $19/bbl.–equivalent to $8/bbl. to $12/bbl. in 1974 dollars.

Levy's projection for 1980 is equal to our estimate at $10/bbl. ($151 billion). The Irving projection for 1980 ($120 billion) is lower than ours at the same assumed price of $10/bbl. in 1974 dollars. Morgan Guaranty's projection is close to ours at $10/bbl. ($143 billion versus $151 billion in 1980). The FNCB projection is considerably lower, reaching only $103 billion in 1980. The World Bank estimate ($170 billion) is greater than our estimate at $11/bbl. ($155 billion), which is the World Bank assumed 1980 price at 1974 dollars.

Projections of OPEC Foreign Capital Accumulation

Table A-7 presents five alternative projections of foreign capital accumulation by OPEC for 1975-1980. These projections are compared to one of our estimates (constant price of $11/bbl.).

Table A-6
Projected OPEC Oil Revenues at Constant and Real Prices, 1976-1980
(Billions of Dollars)

Year	Revenues at Constant 1974 $			Price Index 1974 = 100	Revenues at Current Prices		
	$8	$10	$12		$8	$10	$12
1976	74	91	106	119	88	108	126
1977	77	92	106	128	99	118	136
1978	78	94	105	137	108	129	144
1979	82	95	103	147	121	140	151
1980	85	96	101	157	133	151	159

Table A-7
Alternative Projections of OPEC Foreign Capital Accumulation, 1975-1980
(Billions of Dollars)

Year	World Bank	Morgan	FNCB	Irving	Levy	Author $11
1975	170	137	102	149	122	64
1976		191	139	200		118
1977		231	169	234	264	174
1978		248	188	258		230
1979		235	196	265		292
1980	653	179	189	248	449	352

The highest estimate of foreign capital accumulation is that of the World Bank. Morgan Guaranty and FNCB provide the lowest estimates, though they differ substantially from each other. Irving projections are somewhat higher. Walter Levy's estimates are intermediate. Out estimate derived from Table 10-3 and adjusted to current prices is smaller than Walter Levy's projection. It should be noted that our estimates include only Middle East OPEC members; adding the capital accumulation of the other OPEC members would have increased our estimates by only a small amount, since these countries are not expected to accumulate significant foreign capital.

The differences between our estimates and the Morgan, FNCB, and Irving projections stem partly from differences in projected output (see Tables A-3 and A-4) but mainly from wide differences in projected imports.[a]

---

[a]See Chapters 9 and 10 for the procedure by which net import is estimated.

Bibliography

Abir, M., "The Role of Persian Gulf Oil in Middle East and International Conflicts," unpublished mimeo., 1975.

Adelman, M.A., *The World Petroleum Market,* Baltimore, Johns Hopkins University Press, 1972.

—— et al., "Energy Self-Sufficiency: An Economic Evaluation," *Technology Review,* May 1974, pp. 23-52.

Barnea, A. and Leiber Z., "Dynamic Optimal Pricing to Deter Entry under Constrained Supply," mimeo., Tel-Aviv University, 1975.

Black, F., Jensen, M.C., and Scholes M., "The Capital Asset Pricing Model: Some Empirical Tests," in M.C. Jensen, ed., *Studies in the Theory of Capital Markets,* Praeger, New York, 1972.

Chenery, H.B., "Reconstructing the World Economy," *Foreign Affairs,* January 1975, pp. 242-263.

Dasgupta, P. and Heal G., "The Optimal Depletion of Exhaustible Resources," *Review of Economic Studies,* Symposium, 1974, pp. 3-28.

Erickson, E.W. and Spann, R.M., "Price, Regulation and the Supply of Natural Gas in the U.S.," *Resources for the Future,* Keith Brown, ed., 1972.

Fisher, F.M., *Supply and Cost in the U.S. Petroleum Industry, Two Econometric Studies,* Baltimore, Johns Hopkins Press, 1964.

Hotelling, H., "The Economics of Exhaustible Assets," *Journal of Political Economy,* April, 1931, pp. 137-175.

Houthakker, H.S., "The Price Elasticity of Energy Demand," mimeo, Committee for Economic Development, December, 1973.

—— and Kennedy, M., "Demand for Energy as a Function of Price," mimeo., 1973.

Hudson, E.A., and Jorgenson, D.W., "U.S. Energy Policy and Economic Growth, 1975-2000," *Bell Journal of Economics and Management,* Autumn 1974, pp. 461-514.

Kuenne, R.E., et al., "Intermediate-Term Energy Programs to Protect against Crude-Petroleum Import Interruptions," Institute for Defense Analyses, paper p-1063, September, 1974, Table I.

Levhary, D., and Liviatan, N., "Notes on Hotelling's Economics of Exhausti ble Resources," discussion paper 751, Falk Institute, Jerusalem, 1975.

Levy, W., "World Oil Cooperation or International Chaos," *Foreign Affairs,* July, 1974.

—— "Future OPEC Accumulation of Oil Money," mimeo., New York, June, 1975.

123

Mancke, R.M., "The Long-Run Supply Curve of Crude Oil Produced in the U.S.," *Antitrust Bulletin,* Winter 1970, pp. 727-756.

Nordhaus, W.D., "The Allocation of Energy Resources," *Brookings Papers on Economic Activity,* 3, 1973, pp. 529-570.

Odell, P.R., "The Availability of Indigenous Energy in Western Europe 1973-1998 with Special Reference to Oil and Natural Gas," 1st World Symposium on Energy and Raw Materials, Paris, June, 1974.

Safer, A.E., "Outlook for World Oil: Prices and Petrodollars," *Economic View from One Wall Street,* Irving Trust Company, 20th March, 1975.

Sharpe, W.F., "Mutual Fund Performance," *Journal of Business,* 1966, pp. 119-139.

Sollow, R.M., 'The Economics of Resources or the Resource of Economics," Richard T. Ely Lecture, *American Economic Review,* May 1974, pp. 1-14.

Anvario Español Del Petroleo, Anespe, 1971.

The British Petroleum Company, *BP Statistical Review of the World Oil Industry, 1973,* London, 1974.

B.P. Benzine und Petroleum Aktiengesellschaft Abteilung Volkswirtschaft, *Der Energie Aubenhandel Westeuropaischer,* London, 1969.

Department of Trade and Industry, *United Kingdom Energy Statistics,* London, 1973.

First National City Bank, "Why OPEC's Rocket will Loose its Thrust," *First National City Bank Monthly Review,* June, 1975.

International Bank of Reconstruction and Development, *Prospects for the Developing Countries,* July 8, 1974.

Morgan Guaranty Trust Company of New-York, "Oil, Looking Back and Looking Ahead," *World Financial Markets,* January 21, 1975.

Oil and Gas Journal, December, 1974.

The Petroleum Economist, Vol. XLIII, No. 1, January, 1975.

United Nations, *World Energy Supply,* 1956-1959, Series J, No. 4, New York, 1961.

———— *World Energy Supply,* 1960-1963, Series J, No. 8, New York, 1965.
———— *World Energy Supply,* 1968-1971, Series J, No. 16, New York, 1973.

U.S. Federal Energy Office, "Fossil Fuel and Electricity Demand Forecast by Major Consuming Sectors: Basic Results and Summary Description," June 3, 1974.

About the Author

Haim Ben-Shahar is a professor of economics at Tel-Aviv University and since 1975 President of Tel-Aviv University. Professor Ben-Shahar received his education in economics, both in Israel and in the United States. He received the Ph.D. from New York University Graduate School of Business Administration in 1961. Teaching primarily in areas of managerial economics and corporate finance, Professor Ben-Shahar published several books and numerous articles and research monographs in finance, capital markets, banking, town planning and economic policy. In addition, he has done applied policy research and consulting in Israel and in a number of Western European countries. In 1975, Professor Ben-Shahar headed an Israeli "Public Committee on Tax Reform," whose proposals for an outright change in the Israeli direct tax system were accepted and implemented en toto.

In 1974, Professor Ben-Shahar spent a year at Columbia University Graduate School of Business Administration as a visiting Professor of Finance, at which time he started the research for this book in collaboration with the Hudson Institute.